Compression-Based Methods of Statistical Analysis and Prediction of Time Series

Boris Ryabko • Jaakko Astola • Mikhail Malyutov

Compression-Based Methods of Statistical Analysis and Prediction of Time Series

Boris Ryabko
Inst. of Computational Technologies
Siberian Branch of the Russian Academy
of Sciences
Novosibirsk, Russia

Jaakko Astola
Dept. of Signal Processing
Tampere University of Technology
Tampere, Finland

Mikhail Malyutov
Dept. of Mathematics
Northeastern University
Boston, MA, USA

ISBN 978-3-319-81234-2 ISBN 978-3-319-32253-7 (eBook)
DOI 10.1007/978-3-319-32253-7

Softcover reprint of the hardcover 1st edition 2016

Printed on acid-free paper

This Springer imprint is published by Springer Nature
The registered company is Springer International Publishing AG Switzerland

Preface

Initially, in the 1960s, universal codes were developed for lossless data compression in their storing and transmission. Those codes can efficiently compress sequences generated by stationary and ergodic sources with unknown statistics. In the last twenty years, it was realized that universal codes can be used for solving many important problems of prediction and statistical analysis of time series. This book describes recent results in this area.

The first chapter of this book is mainly devoted to the application of universal codes to prediction and statistical analysis of time series. The applications of suggested statistical methods to cryptography are quite numerous, so they are described separately in Chap. 2. These two chapters were written by B. Ryabko and J. Astola.

The third chapter presents a sketch of the theory behind many applications of a simplified homogeneity test between literary texts based on universal compressors. In particular, this test can be used for authorship attribution if training texts written by different candidates are available. This chapter was written by M. Malyutov.

Boris Ryabko is partly supported by Russian Science Foundation, grant 14-14-00603 and by Russian Foundation for Basic Research, grant 15-29-07932.

Novosibirsk, Russia — B. Ryabko
Tampere, Finland — J. Astola
Boston, MA, USA — M. Malyutov
August, 2015

Contents

Chapter 1
Statistical Methods Based on Universal Codes

1.1 Introduction

We show how universal codes can be used for solving some of the most important statistical problems for time series. By definition, a universal code (or a universal lossless data compressor) can compress any sequence generated by a stationary and ergodic source asymptotically to the Shannon entropy, which, in turn, is the best achievable ratio for lossless data compressors.

First we show how universal codes can be used for solving some problems of time series analysis and then apply obtained methods to several real problems.

We consider finite-alphabet and real-valued time series and the following problems: estimation of the limiting probabilities for finite-alphabet time series and estimation of the density for real-valued time series; the on-line prediction, regression, and classification (or problems with side information) for both types of time series; and two problems of hypothesis testing (namely, goodness-of-fit testing, or identity testing and testing of serial independence). It is important to note that all problems are considered in the framework of classical mathematical statistics and, on the other hand, everyday methods of data compression (or archivers) can be used as a tool for estimation and testing.

It turns out that quite often the suggested methods and tests are more powerful than known ones when they are applied in practice. The applications are intended to show practical efficiency of the obtained methods.

Since C. Shannon published the paper "A mathematical theory of communication" [46], the ideas and results of Information Theory have played an important role in cryptography [47], mathematical statistics [3, 8, 26], and many other fields [6, 7], which are far from telecommunication. Universal coding, which is a part of Information Theory, also has been efficiently applied in many fields since its discovery [13, 19]. Thus, application of results of universal coding, initiated in 1988 [33], created a new approach to prediction [1, 20, 27, 28]. Maybe the most unexpected application of data compression ideas arises in experiments that show

B. Ryabko et al., *Compression-Based Methods of Statistical Analysis and Prediction of Time Series*, DOI 10.1007/978-3-319-32253-7_1

that some ant species are capable of compressing messages and are capable of adding and subtracting small numbers [41, 42].

In this chapter we describe a new approach to estimation, prediction and hypothesis testing for time series, which was suggested recently [33, 36, 37]. This approach is based on ideas of universal coding (or universal data compression). We would like to emphasize that everyday methods of data compression (or archivers) can be directly used as a tool for estimation and hypothesis testing. It is important to note that modern archivers (like *zip*, *arj*, *rar*, etc.) are based on deep theoretical results of source coding theory [10, 21, 25, 31, 45] and have shown their high efficiency in practice because archivers can find many kinds of latent regularities and use them for compression.

The outline of this chapter is as follows. The next section contains some necessary definitions and facts about predictors, codes, hypothesis testing and a description of one universal code. Sections 1.3 and 1.4 are devoted to problems of estimation and hypothesis testing, correspondingly, for the case of finite-alphabet time series. The case of infinite alphabets is considered in Sect. 1.6. All proofs are given in Sect. 1.8, but some intuitive indications are given in the body of the chapter.

1.2 Definitions and Statements of the Problems

1.2.1 Estimation and Prediction for I.I.D. Sources

First we consider a source with unknown statistics which generates sequences $x_1x_2\cdots$ of letters from some set (or alphabet) A. It will be convenient now to describe briefly the prediction problem. Let the source generate a message $x_1\ldots x_{t-1}x_t$, $x_i \in A$ for all i; the following letter x_{t+1} needs to be predicted. This problem can be traced back to Laplace [11, 29] who considered the problem of estimation of the probability that the sun will rise tomorrow, given that it has risen every day since Creation. In our notation the alphabet A contains two letters, 0 ("the sun rises") and 1 ("the sun does not rise"); t is the number of days since Creation, $x_1\ldots x_{t-1}x_t = 00\ldots 0$.

Laplace suggested the following predictor:

$$L_0(a|x_1\cdots x_t) = (\nu_{x_1\cdots x_t}(a) + 1)/(t + |A|), \tag{1.1}$$

where $\nu_{x_1\cdots x_t}(a)$ denotes the count of letter a occurring in the word $x_1\ldots x_{t-1}x_t$. It is important to note that the predicted probabilities cannot be equal to zero even through a certain letter did not occur in the word $x_1\ldots x_{t-1}x_t$.

Example Let $A = \{0, 1\}$, $x_1\ldots x_5 = 01010$; then the Laplace prediction is as follows: $L_0(x_6 = 0|x_1\ldots x_5 = 01010) = (3+1)/(5+2) = 4/7$, $L_0(x_6 = 1|x_1\ldots x_5 = 01010) = (2+1)/(5+2) = 3/7$. In other words, 3/7 and 4/7 are estimations of the unknown probabilities $P(x_{t+1} = 0|x_1\ldots x_t = 01010)$ and

$P(x_{t+1} = 1|x_1 \ldots x_t = 01010)$. (In what follows we will use the shorter notation: $P(0|01010)$ and $P(1|01010)$.)

We can see that Laplace considered prediction as a set of estimations of unknown (conditional) probabilities. This approach to the problem of prediction was developed in 1988 [33] and now is often called on-line prediction or universal prediction [1, 20, 27, 28, 40, 43]. As we mentioned above, it seems natural to consider conditional probabilities to be the best predictions, because they contain all information about the future behavior of the stochastic process. Moreover, this approach is deeply connected with game-theoretical interpretation of prediction [18, 35] and, in fact, all obtained results can be easily transferred from one model to the other.

Any predictor γ defines a measure (or an estimation of probability) by the following equation:

$$\gamma(x_1 \ldots x_t) = \prod_{i=1}^{t} \gamma(x_i|x_1 \ldots x_{i-1}). \tag{1.2}$$

And, vice versa, any measure γ (or estimation of probability) defines a predictor:

$$\gamma(x_i|x_1 \ldots x_{i-1}) = \gamma(x_1 \ldots x_{i-1}x_i)/\gamma(x_1 \ldots x_{i-1}). \tag{1.3}$$

Example Let us apply the Laplace predictor for estimation of probabilities of the sequences 01010 and 010101. From (1.2) we obtain $L_0(01010) = \frac{1}{2}\frac{1}{3}\frac{2}{4}\frac{2}{5}\frac{3}{6} = \frac{1}{60}$, $L_0(010101) = \frac{1}{60}\frac{3}{7} = \frac{1}{140}$. Vice versa, if for some measure (or a probability estimation) χ we have $\chi(01010) = \frac{1}{60}$ and $\chi(010101) = \frac{1}{140}$, then we obtain from (1.3) the following prediction, or the estimation of the conditional probability, $\chi(1|01010) = \frac{1/140}{1/60} = \frac{3}{7}$.

Now we concretize the class of stochastic processes which will be considered. Generally speaking, we will deal with so-called stationary and ergodic time series (or sources), whose definition will be given later; but now we consider what may be the simplest class of such processes, which are called i.i.d. sources. By definition, they generate independent and identically distributed random variables from some set A. In our case A will be either some alphabet or a real-valued interval.

The next natural question is how to measure the errors of prediction and estimation of probability. Mainly we will measure these errors by the Kullback-Leibler (KL) divergence, which is defined by

$$D(P, Q) = \sum_{a \in A} P(a) \log \frac{P(a)}{Q(a)}, \tag{1.4}$$

where $P(a)$ and $Q(a)$ are probability distributions over an alphabet A (here and below $\log \equiv \log_2$ and $0 \log 0 = 0$). The probability distribution $P(a)$ can be considered as unknown whereas $Q(a)$ is its estimation. It is well-known that for any distributions P and Q the KL divergence is nonnegative and equals 0 if and only if $P(a) = Q(a)$ for all a [14]. So, if the estimation Q is equal to P, the error is 0; otherwise the error is a positive number.

The KL divergence is connected with the so-called variation distance

$$||P - Q|| = \sum_{a \in A} |P(a) - Q(a)|,$$

via the the following inequality (Pinsker's inequality):

$$\sum_{a \in A} P(a) \log \frac{P(a)}{Q(a)} \geq \frac{\log e}{2} ||P - Q||^2. \tag{1.5}$$

Let γ be a predictor, i.e. an estimation of an unknown conditional probability, and $x_1 \cdots x_t$ be a sequence of letters created by an unknown source P. The KL divergence between P and the predictor γ is equal to

$$\rho_{\gamma,P}(x_1 \cdots x_t) = \sum_{a \in A} P(a|x_1 \cdots x_t) \log \frac{P(a|x_1 \cdots x_t)}{\gamma(a|x_1 \cdots x_t)}. \tag{1.6}$$

For fixed t it is a random variable, because $x_1, x_2, \cdots, x_t$ are random variables. We define the average error at time t by

$$\rho^t(P\|\gamma) = E\left(\rho_{\gamma,P}(\cdot)\right) = \sum_{x_1 \cdots x_t \in A^t} P(x_1 \cdots x_t)\ \rho_{\gamma,P}(x_1 \cdots x_t) \tag{1.7}$$

$$= \sum_{x_1 \cdots x_t \in A^t} P(x_1 \cdots x_t) \sum_{a \in A} P(a|x_1 \cdots x_t) \log \frac{P(a|x_1 \cdots x_t)}{\gamma(a|x_1 \cdots x_t)}.$$

Analogously, if $\gamma()$ is an estimation of a probability distribution we define the errors *per letter* as follows:

$$\bar{\rho}_{\gamma,P}(x_1 \dots x_t) = t^{-1} \left(\log(P(x_1 \dots x_t)/\gamma(x_1 \dots x_t))\right) \tag{1.8}$$

and

$$\bar{\rho}^t(P\|\gamma) = t^{-1} \sum_{x_1 \dots x_t \in A^t} P(x_1 \dots x_t) \log(P(x_1 \dots x_t)/\gamma(x_1 \dots x_t)), \tag{1.9}$$

where, as before, $\gamma(x_1 \dots x_t) = \prod_{i=1}^t \gamma(x_i|x_1 \dots x_{i-1})$. (Here and below we denote by A^t and A^* the set of all words of length t over A and the set of all finite words over A correspondingly: $A^* = \bigcup_{i=1}^\infty A^i$.)

Claim 1.1 ([33]) For any i.i.d. source P generating letters from an alphabet A and an integer t the average error (1.7) of the Laplace predictor and the average error of the Laplace estimator are upper bounded as follows:

$$\rho^t(P\|L_0) \le ((|A|-1)\log e)/(t+1), \tag{1.10}$$

$$\bar{\rho}^t(P\|L_0) \le (|A|-1)\log t/t + O(1/t), \tag{1.11}$$

where $e \simeq 2.718$ is the Euler number.

So, we can see that the average error of the Laplace predictor goes to zero for any i.i.d. source P when the length t of the sample $x_1 \cdots x_t$ tends to infinity. Such methods are called universal, because the error goes to zero for any source, or process. In this case they are universal for the set of all i.i.d. sources generating letters from the finite alphabet A, but later we consider universal estimators for the set of stationary and ergodic sources. It is worth noting that the first universal code for which the estimation (1.11) is valid was suggested independently by Fitingof [13] and Kolmogorov [22] in 1966.

The value

$$\bar{\rho}^t(P\|\gamma) = t^{-1} \sum_{x_1 \dots x_t \in A^t} P(x_1 \dots x_t) \log(P(x_1 \dots x_t)/\gamma(x_1 \dots x_t))$$

has one more interpretation connected with data compression. Now we consider the main idea whereas the more formal definitions will be given later. First we recall the definition of the Shannon entropy $h_0(P)$ for an i.i.d. source P,

$$h_0(P) = -\sum_{a \in A} P(a) \log P(a). \tag{1.12}$$

It is easy to see that $t^{-1} \sum_{x_1 \dots x_t \in A^t} P(x_1 \dots x_t) \log(P(x_1 \dots x_t)) = -h_0(P)$ for the i.i.d. source. Hence, we can represent the average error $\bar{\rho}^t(P\|\gamma)$ in (1.9) as

$$\bar{\rho}^t(P\|\gamma) = t^{-1} \sum_{x_1 \dots x_t \in A^t} P(x_1 \dots x_t) \log(1/\gamma(x_1 \dots x_t)) - h_0(P).$$

More formal and general consideration of universal codes will be given later, but here we briefly show how estimations and codes are connected. The point is that one can construct a code with codelength $\gamma_{code}(a|x_1 \cdots x_t) \approx -\log_2 \gamma(a|x_1 \cdots x_n)$ for any letter $a \in A$ (since Shannon's original research, it has been well known that, using block codes with large block length or more modern methods of arithmetic coding [30], the approximation may be as accurate as you like). If one knows the real distribution P, one can base coding on the true distribution P and not on the prediction γ. The difference in performance measured by average code length is given by

$$\sum_{a \in A} P(a|x_1 \cdots x_t)(-\log_2 \gamma(a|x_1 \cdots x_t)) - \sum_{a \in A} P(a|x_1 \cdots x_t)(-\log_2 P(a|x_1 \cdots x_t))$$
$$= \sum_{a \in A} P(a|x_1 \cdots x_t) \log_2 \frac{P(a|x_1 \cdots x_t)}{\gamma(a|x_1 \cdots x_t)}.$$

Thus this excess is exactly the error defined above (1.6). Analogously, if we encode the sequence $x_1 \dots x_t$ based on a predictor γ the redundancy per letter is defined by (1.8) and (1.9). So, from a mathematical point of view, the estimation of the limiting probabilities and universal coding are identical. But $-\log \gamma(x_1 \dots x_t)$ and $-\log P(x_1 \dots x_t)$ have a very natural interpretation. The first value is a code word length (in bits), if the "code" γ is applied for compressing the word $x_1 \dots x_t$, and the second one is the minimal possible codeword length. The difference is the redundancy of the code and, at the same time, the error of the predictor. It is worth noting that there are many other deep interrelations between the universal coding, prediction and estimation [31, 33].

We can see from the claim and the Pinsker inequality (1.5) that the variation distance of the Laplace predictor and estimator goes to zero, too. Moreover, it can be easily shown that the error (1.6) (and the corresponding variation distance) goes to zero with probability 1 when t goes to infinity. (Informally, it means that the error (1.6) goes to zero for almost all sequences $x_1 \cdots x_t$ according to the measure P.) Obviously, such properties are very desirable for any predictor and for larger classes of sources, like Markov and stationary ergodic (they will be briefly defined in the next subsection). However, it is proven [33] that such predictors do not exist for the class of all stationary and ergodic sources (generating letters from a given finite alphabet). More precisely, if, for example, the alphabet has two letters, then for any predictor γ and for any $\delta > 0$ there exists a source P such that with probability 1 $\rho_{\gamma,P}(x_1 \cdots x_t) \geq 1/2 - \delta$ infinitely often when $t \to \infty$. In other words, the error of any predictor may not go to 0 if the predictor is applied to an arbitrary stationary and ergodic source. That is why it is difficult to use (1.6) and (1.7) to compare different predictors. On the other hand, it is shown [33] that there exists a predictor R such that the following Cesáro average $t^{-1} \sum_{i=1}^{t} \rho_{R,P}(x_1 \cdots x_i)$ goes to 0 (with probability 1) for any stationary and ergodic source P, where t goes to infinity. (This predictor will be described in the next subsection.) That is why we will focus our attention on such averages. From the definitions (1.6) and (1.7) and properties of the logarithm we can see that for any probability distribution γ,

$$t^{-1} \sum_{i=1}^{t} \rho_{\gamma,P}(x_1 \cdots x_i) = t^{-1} \left(\log(P(x_1 \dots x_t)/\gamma(x_1 \dots x_t)\right),$$

$$t^{-1} \sum_{i=1}^{t} \rho^i(P\|\gamma) = t^{-1} \sum_{x_1 \dots x_t \in A^t} P(x_1 \dots x_t) \log(P(x_1 \dots x_t)/\gamma(x_1 \dots x_t)).$$

Taking into account these equations, we can see from the definitions (1.8) and (1.9) that the Cesáro averages of the prediction errors (1.6) and (1.7) are equal to the errors of estimation of limiting probabilities (1.8) and (1.9). That is why we will use values (1.8) and (1.9) as the main measures of the precision throughout the chapter.

A natural problem is to find a predictor and an estimator of the limiting probabilities whose average error (1.9) is minimal for the set of i.i.d. sources. This problem was considered and solved by Krichevsky [24]. He suggested the following predictor:

$$K_0(a|x_1\cdots x_t) = (\nu_{x_1\cdots x_t}(a) + 1/2)/(t + |A|/2), \tag{1.13}$$

where, as before, $\nu_{x_1\cdots x_t}(a)$ is the number of occurrences of the letter a in the word $x_1 \dots x_t$. We can see that the Krichevsky predictor is quite close to Laplace's (1.1).

Example Let $A = \{0, 1\}$, $x_1 \dots x_5 = 01010$. Then $K_0(x_6 = 0|01010) = (3 + 1/2)/(5 + 1) = 7/12$, $K_0(x_6 = 1|01010) = (2 + 1/2)/(5 + 1) = 5/12$ and $K_0(01010) = \frac{1}{2}\frac{1}{4}\frac{1}{2}\frac{3}{8}\frac{1}{2} = \frac{3}{256}$.

The Krichevsky measure K_0 can be represented as follows:

$$K_0(x_1 \dots x_t) = \prod_{i=1}^{t} \frac{\nu_{x_1\dots x_{i-1}}(x_i) + 1/2}{i - 1 + |A|/2} = \frac{\prod_{a\in A}(\prod_{j=1}^{\nu_{x_1\dots x_t}(a)}(j - 1/2))}{\prod_{i=0}^{t-1}(i + |A|/2)}. \tag{1.14}$$

It is known that

$$(r + 1/2)((r + 1) + 1/2)\dots(s - 1/2) = \frac{\Gamma(s + 1/2)}{\Gamma(r + 1/2)}, \tag{1.15}$$

where $\Gamma()$ is the gamma function [23]. So, (1.14) can be presented as follows:

$$K_0(x_1 \dots x_t) = \frac{\prod_{a\in A}(\Gamma(\nu_{x_1\dots x_t}(a) + 1/2)\,/\,\Gamma(1/2))}{\Gamma(t + |A|/2)\,/\,\Gamma(|A|/2)}. \tag{1.16}$$

The following claim shows that the error of the Krichevsky estimator is half of Laplace's.

Claim 1.2 For any i.i.d. source P generating letters from a finite alphabet A the average error (1.9) of the estimator K_0 is upper bounded as follows:

$$\bar{\rho}_t(K_0, P) \equiv t^{-1} \sum_{x_1\dots x_t\in A^t} P(x_1 \dots x_t) \log(P(x_1 \dots x_t)/K_0(x_1 \dots x_t)) \equiv \tag{1.17}$$

$$t^{-1} \sum_{x_1\dots x_t\in A^t} P(x_1 \dots x_t) \log(1/K_0(x_1 \dots x_t)) - h_0(p) \le ((|A| - 1)\log t + C)/(2t),$$

where C is a constant.

Moreover, in a certain sense this average error is minimal: it is shown by Krichevsky [24] that for any predictor γ there exists such a source P^* that

$$\bar{\rho}_t(\gamma, P^*) \geq ((|A| - 1) \log t + C')/(2t).$$

Hence, the bound $((|A| - 1) \log t + C)/(2t)$ cannot be reduced and the Krichevsky estimator is the best (up to $O(1/t)$) if the error is measured by the KL divergence ρ.

1.2.2 Consistent Estimations and On-Line Predictors for Markov and Stationary Ergodic Processes

Now we briefly describe consistent estimations of unknown probabilities and efficient on-line predictors for general stochastic processes (or sources of information).

First we give a formal definition of stationary ergodic processes. The time shift T on A^∞ is defined as $T(x_1, x_2, x_3, \dots) = (x_2, x_3, \dots)$. A process P is called stationary if it is T-invariant: $P(T^{-1}B) = P(B)$ for every Borel set $B \subset A^\infty$. A stationary process is called ergodic if every T-invariant set has probability 0 or 1: $P(B) = 0$ or 1 whenever $T^{-1}B = B$ [4, 14].

We denote by $M_\infty(A)$ the set of all stationary and ergodic sources and let $M_0(A) \subset M_\infty(A)$ be the set of all i.i.d. processes. We denote by $M_m(A) \subset M_\infty(A)$ the set of Markov sources of order (or with memory, or connectivity) not larger than m, $m \geq 0$. By definition $\mu \in M_m(A)$ if

$$\mu(x_{t+1} = a_{i_1} | x_t = a_{i_2}, x_{t-1} = a_{i_3}, \dots, x_{t-m+1} = a_{i_{m+1}}, \dots) \tag{1.18}$$
$$= \mu(x_{t+1} = a_{i_1} | x_t = a_{i_2}, x_{t-1} = a_{i_3}, \dots, x_{t-m+1} = a_{i_{m+1}})$$

for all $t \geq m$ and $a_{i_1}, a_{i_2}, \dots \in A$. Let $M^*(A) = \bigcup_{i=0}^{\infty} M_i(A)$ be the set of all finite-order sources.

The Laplace and Krichevsky predictors can be extended to general Markov processes. The trick is to view a Markov source $p \in M_m(A)$ as resulting from $|A|^m$ i.i.d. sources. We illustrate this idea by an example. Assume that $A = \{O, I\}$ and $m = 2$ and suppose that the source $p \in M_2(A)$ has generated the sequence

$$OOIOIIOOIIIOIO.$$

We represent this sequence by the following four subsequences:

$$**I*****I*****,$$

$$***O*I***I***O,$$

$$****I**O****I*,$$

$$******O***IO**.$$

These four subsequences contain letters which follow OO, OI, IO and II, respectively. By definition, $p \in M_m(A)$ if $p(a|x_t\cdots x_1) = p(a|x_t\cdots x_{t-m+1})$, for all $0 < m \le t$, all $a \in A$ and all $x_1\cdots x_t \in A^t$. Therefore, each of the four generated subsequences may be considered to be generated by an i.i.d. source. Further, it is possible to reconstruct the original sequence if we know the four $(= |A|^m)$ subsequences and the two $(= m)$ first letters of the original sequence.

Any predictor γ for i.i.d. sources can be applied to Markov sources. Indeed, in order to predict, it is enough to store in the memory $|A|^m$ sequences, one corresponding to each word in A^m. Thus, in the example, the letter x_3 which follows OO is predicted based on the i.i.d. method γ corresponding to the x_1x_2-subsequence $(= OO)$; then x_4 is predicted based on the i.i.d. method corresponding to x_2x_3, i.e., to the OI-subsequence, and so forth. When this scheme is applied along with either L_0 or K_0 we denote the obtained predictors as L_m and K_m, correspondingly, and define the probabilities for the first m letters as follows: $L_m(x_1) = L_m(x_2) = \ldots = L_m(x_m) = 1/|A|$, $K_m(x_1) = K_m(x_2) = \ldots = K_m(x_m) = 1/|A|$. For example, having taken into account (1.16), we can present the Krichevsky predictors for $M_m(A)$ as follows:

$$K_m(x_1\ldots x_t) = \begin{cases} \frac{1}{|A|^t}, & \text{if } t \le m, \\ \frac{1}{|A|^m} \prod_{v\in A^m} \frac{\prod_{a\in A}((\Gamma(\nu_x(va)+1/2)\,/\,\Gamma(1/2))}{(\Gamma(\bar{\nu}_x(v)+|A|/2)\,/\,\Gamma(|A|/2))}, & \text{if } t > m, \end{cases} \tag{1.19}$$

where $\bar{\nu}_x(v) = \sum_{a\in A} \nu_x(va)$, $x = x_1\ldots x_t$. It is worth noting that the representation (1.14) can be more convenient for carrying out calculations if t is small.

Example For the word $OOIOIIOOIIIOIO$ considered in the previous example, we obtain $K_2(OOIOIIOOIIIOIO) = 2^{-2}\ \frac{1}{2}\frac{3}{4}\ \frac{1}{2}\frac{1}{4}\frac{1}{2}\frac{3}{8}\ \frac{1}{2}\frac{1}{4}\frac{1}{2}\ \frac{1}{2}\frac{1}{4}\frac{1}{2}$. Here groups of multipliers correspond to subsequences II, $OIIO$, IOI, OIO.

In order to estimate the error of the Krichevsky predictor K_m we need a general definition of the Shannon entropy. Let P be a stationary and ergodic source generating letters from a finite alphabet A. The m-order (conditional) Shannon entropy and the limiting Shannon entropy are defined as follows:

$$h_m(P) = \sum_{v\in A^m} P(v) \sum_{a\in A} P(a/v) \log P(a/v), \qquad h_\infty(\tau) = \lim_{m\to\infty} h_m(P). \tag{1.20}$$

(If $m = 0$ we obtain the definition (1.12).) It is also known that for any m

$$h_\infty(P) \le h_m(P)\,, \tag{1.21}$$

see [4, 14].

Claim 1.3 For any stationary and ergodic source P generating letters from a finite alphabet A, the average error of the Krichevsky predictor K_m is upper bounded as follows:

$$-t^{-1} \sum_{x_1 \dots x_t \in A^t} P(x_1 \dots x_t) \log(K_m(x_1 \dots x_t)) - h_m(P) \le \frac{|A|^m(|A|-1)\log t + C}{2t}, \tag{1.22}$$

where C is a constant.

The following so-called empirical Shannon entropy, which is an estimation of the entropy (1.20), will play a key role in the hypothesis testing. It will be convenient to consider its definition here, because this notation will be used in the proof of the next claims. Let $v = v_1 \dots v_k$ and $x = x_1 x_2 \dots x_t$ be words from A^*. Denote the rate of a word v occurring in the sequence $x = x_1 x_2 \dots x_k$, $x_2 x_3 \dots x_{k+1}$, $x_3 x_4 \dots x_{k+2}$, $\dots$, $x_{t-k+1} \dots x_t$ as $\nu_x(v)$. For example, if $x = 000100$ and $v = 00$, then $\nu_x(00) = 3$. For any $0 \le k < t$ the empirical Shannon entropy of order k is defined as follows:

$$h_k^*(x) = - \sum_{v \in A^k} \frac{\bar{\nu}_x(v)}{(t-k)} \sum_{a \in A} \frac{\nu_x(va)}{\bar{\nu}_x(v)} \log \frac{\nu_x(va)}{\bar{\nu}_x(v)}, \tag{1.23}$$

where $x = x_1 \dots x_t$, $\bar{\nu}_x(v) = \sum_{a \in A} \nu_x(va)$. In particular, if $k = 0$, we obtain $h_0^*(x) = -t^{-1} \sum_{a \in A} \nu_x(a) \log(\nu_x(a)/t)$.

Let us define the measure R, which, in fact, is a consistent estimator of probabilities for the class of all stationary and ergodic processes with a finite alphabet. First we define a probability distribution $\{\omega = \omega_1, \omega_2, \dots\}$ on integers $\{1, 2, \dots\}$ by

$$\omega_1 = 1 - 1/\log 3, \ \dots \ , \ \omega_i = 1/\log(i+1) - 1/\log(i+2), \ \dots \ . \tag{1.24}$$

(In what follows we will use this distribution, but results described below are obviously true for any distribution with nonzero probabilities.) The measure R is defined as follows:

$$R(x_1 \dots x_t) = \sum_{i=0}^{\infty} \omega_{i+1} \, K_i(x_1 \dots x_t). \tag{1.25}$$

It is worth noting that this construction can be applied to the Laplace measure (if we use L_i instead of K_i) and any other family of measures.

Example Let us calculate $R(00), \dots, R(11)$. From (1.14) and (1.19) we obtain:

$$K_0(00) = K_0(11) = \frac{1/2}{1} \frac{3/2}{1+1} = 3/8, \ K_0(01) = K_0(10) = \frac{1/2}{1+0} \frac{1/2}{1+1} = 1/8,$$

$$K_i(00) = K_i(01) = K_i(10) = K_i(11) = 1/4; \ , \ i \ge 1.$$

Having taken into account the definitions of ω_i (1.24) and the measure R (1.25), we can calculate $R(z_1 z_2)$ as follows:

$$R(00) = \omega_1 K_0(00) + \omega_2 K_1(00) + \ldots = (1 - 1/\log 3)\, 3/8 + (1/\log 3 - 1/\log 4)\, 1/4 +$$
$$(1/\log 4 - 1/\log 5)\, 1/4 + \ldots = (1 - 1/\log 3)\ 3/8 + (1/\log 3)\ 1/4 \approx 0.296.$$

Analogously, $R(01) = R(10) \approx 0.204, R(11) \approx 0.296$.

The main properties of the measure R are connected with the Shannon entropy (1.20).

Theorem 1.1 ([33]) *For any stationary and ergodic source P the following equalities are valid:*

$$i)\ \lim_{t\to\infty} \frac{1}{t} \log(1/R(x_1 \cdots x_t)) \ = \ h_\infty(P)$$

with probability 1,

$$ii)\ \lim_{t\to\infty} \frac{1}{t} \sum_{u\in A^t} P(u) \log(1/R(u)) \ = \ h_\infty(P).$$

So, if one uses the measure R for data compression in such a way that the codeword length of the sequence $x_1 \cdots x_t$ is (approximately) equal to $\log(1/R(x_1 \cdots x_t))$ bits, he or she obtains the best achievable data compression ratio $h_\infty(P)$ per letter. On the other hand, we know that the redundancy of a universal code and the error of the corresponding predictor are equal. Hence, if one uses the measure R for estimation and/or prediction, the error (per letter) will go to zero.

1.2.3 Hypothesis Testing

Here we briefly describe the main notions of hypothesis testing and the two particular problems considered below. A statistical test is formulated to test a specific null hypothesis (H_0). Associated with this null hypothesis is the alternative hypothesis (H_1) [33]. For example, we will consider the two following problems: goodness-of-fit testing (or identity testing) and testing of serial independence. Both problems are well known in mathematical statistics and there is extensive literature dealing with their nonparametric testing [2, 8, 9, 12].

The goodness-of-fit testing is described as follows: a hypothesis H_0^{id} is that the source has a particular distribution π and the alternative hypothesis H_1^{id} that the sequence is generated by a stationary and ergodic source which differs from the source under H_0^{id}. One particular case, mentioned in the Introduction, is when the source alphabet A is $\{0, 1\}$ and the main hypothesis H_0^{id} is that a bit sequence is generated by the Bernoulli i.i.d. source with equal probabilities of 0's and 1's. In all cases, the testing should be based on a sample $x_1 \dots x_t$ generated by the source.

The second problem is as follows: the null hypothesis H_0^{SI} is that the source is Markovian of order not larger than m ($m \geq 0$), and the alternative hypothesis H_1^{SI} is that the sequence is generated by a stationary and ergodic source which differs from the source under H_0^{SI}. In particular, if $m = 0$, this is the problem of testing for independence of time series.

For each applied test, a decision is derived that accepts or rejects the null hypothesis. During the test, a test statistic value is computed on the data (the sequence being tested). This test statistic value is compared to the critical value. If the test statistic value exceeds the critical value, the null hypothesis is rejected. Otherwise, the null hypothesis is accepted. So, statistical hypothesis testing is a conclusion-generation procedure that has two possible outcomes: either accept H_0 or accept H_1.

Errors of the two following types are possible: The Type I error occurs if H_0 is true but the test accepts H_1 and, vice versa, the Type II error occurs if H_1 is true but the test accepts H_0. The probability of Type I error is often called the level of significance of the test. This probability can be set prior to the testing and is denoted by α. For a test, α is the probability that the test will say that H_0 is not true when it really is true. Common values of α are about 0.01. The probabilities of Type I and Type II errors are related to each other and to the size n of the tested sequence in such a way that if two of them are specified, the third value is automatically determined. Practitioners usually select a sample size n and a value for the probability of the Type I error [33].

1.2.4 Codes

We briefly describe the main definitions and properties (without proofs) of lossless codes, or methods of (lossless) data compression. A data compression method (or code) φ is defined as a set of mappings φ_n such that $\varphi_n : A^n \to \{0,1\}^*$, $n = 1, 2, \dots$, and for each pair of different words $x, y \in A^n$ $\varphi_n(x) \neq \varphi_n(y)$. It is also required that each sequence $\varphi_n(u_1)\varphi_n(u_2)\dots\varphi_n(u_r)$, $r \geq 1$, of encoded words from the set $A^n, n \geq 1$, be decodable into $u_1 u_2 \dots u_r$. Such codes are called uniquely decodable. For example, let $A = \{a, b\}$; the code $\psi_1(a) = 0, \psi_1(b) = 00$, obviously, is not uniquely decodable. In what follows we call uniquely decodable codes just "codes". It is well known that if φ is a code then the lengths of the codewords satisfy the following inequality (Kraft's inequality) [14]: $\Sigma_{u \in A^n} 2^{-|\varphi_n(u)|} \leq 1$. It will be convenient to reformulate this property as follows:

Claim 1.4 Let φ be a code over an alphabet A. Then for any integer n there exists a measure μ_φ on A^n such that

$$- \log \mu_\varphi(u) \leq |\varphi(u)| \tag{1.26}$$

for any u from A^n.

(Obviously, this claim is true for the measure $\mu_\varphi(u) = \frac{2^{-|\varphi(u)|}}{\Sigma_{u \in A^n} 2^{-|\varphi(u)|}}$.)

It was mentioned above that, in a certain sense, the opposite claim is true, too. Namely, for any probability measure μ defined on $A^n, n \geq 1$, there exists a code φ_μ such that

$$|\varphi_\mu(u)| = -\log \mu(u). \tag{1.27}$$

(More precisely, for any $\varepsilon > 0$ one can construct a code φ^*_μ, such that $|\varphi^*_\mu(u)| < -\log \mu(u) + \varepsilon$ for any $u \in A^n$. Such a code can be constructed by applying a so-called arithmetic coding [30] .) For example, for the above described measure R we can construct a code R_{code} such that

$$|R_{code}(u)| = -\log R(u). \tag{1.28}$$

As we mentioned above there exist universal codes. For their description we recall that sequences $x_1 \dots x_t$, generated by a source P, can be "compressed" to the length $-\log P(x_1 \dots x_t)$ bits (see (1.27)) and, on the other hand, for any source P there is no code ψ for which the average codeword length ($\Sigma_{u \in A^t} P(u)|\psi(u)|$) is less than $-\Sigma_{u \in A^t} P(u) \log P(u)$. Universal codes can reach the lower bound $-\log P(x_1 \dots x_t)$ asymptotically for any stationary and ergodic source P on average and with probability 1. The formal definition is as follows: a code U is universal if for any stationary and ergodic source P the following equalities are valid:

$$\lim_{t \to \infty} |U(x_1 \dots x_t)|/t = h_\infty(P) \tag{1.29}$$

with probability 1, and

$$\lim_{t \to \infty} E(|U(x_1 \dots x_t)|)/t = h_\infty(P), \tag{1.30}$$

where $E(f)$ is the expected value of f, and $h_\infty(P)$ is the Shannon entropy of P; see (1.21). So, informally speaking, a universal code estimates the probability characteristics of a source and uses them for efficient "compression".

In this chapter we mainly consider finite-alphabet and real-valued sources, but sources with countable alphabet also were considered by many authors [15, 17, 19, 38, 39]. In particular, it is shown that, for infinite alphabet, without any condition on the source distribution it is impossible to have universal source code and/or universal predictor, i.e., such a predictor whose average error goes to zero, when the length of a sequence goes to infinity. On the other hand, there are some necessary and sufficient conditions for existence of universal codes and predictors [15, 19, 38].

1.3 Finite Alphabet Processes

1.3.1 The Estimation of (Limiting) Probabilities

The following theorem shows how universal codes can be applied for probability estimation.

Theorem 1.2 *Let U be a universal code and*

$$\mu_U(u) = 2^{-|U(u)|} / \Sigma_{v \in A^{|u|}} 2^{-|U(v)|}. \tag{1.31}$$

Then, for any stationary and ergodic source P the following equalities are valid:

$$i)\ \lim_{t\to\infty} \frac{1}{t}(-\log P(x_1 \cdots x_t) - (-\log \mu_U(x_1 \cdots x_t))) = 0$$

with probability 1,

$$ii)\ \lim_{t\to\infty} \frac{1}{t} \sum_{u \in A^t} P(u) \log(P(u)/\mu_U(u)) = 0.$$

The informal outline of the proof is as follows: $\frac{-1}{t}(\log P(x_1 \cdots x_t)$ and $\frac{-1}{t}$ (log $\mu_U(x_1 \cdots x_t))$ goes to the Shannon entropy $h_\infty(P)$; that is why the difference is 0.

So, we can see that, in a certain sense, the measure μ_U is a consistent nonparametric estimation of the (unknown) measure P.

Nowadays there are many efficient universal codes (and universal predictors connected with them) which can be applied to estimation. For example, the above described measure R is based on a universal code [32, 33] and can be applied for probability estimation. More precisely, Theorem 1.2 (and the following theorems) are true for R, if we replace μ_U by R.

It is important to note that the measure R has some additional properties, which can be useful for applications. The following theorem describes these properties (whereas all other theorems are valid for all universal codes and corresponding measures, including the measure R).

Theorem 1.3 ([32, 33]) *For any Markov process P with memory k:*

i) The error of the probability estimator, which is based on the measure R, is upper-bounded as follows:

$$\frac{1}{t}\sum_{u\in A^t} P(u)\log(P(u)/R(u)) \le \frac{(|A|-1)|A|^k \log t}{2t} + O\left(\frac{1}{t}\right).$$

ii) The error of R is asymptotically minimal in the following sense: for any measure μ there exists a k-memory Markov process p_μ such that

$$\frac{1}{t}\sum_{u\in A^t} p_\mu(u)\log(p_\mu(u)/\mu(u)) \ge \frac{(|A|-1)\,|A|^k \log t}{2t} + O\left(\frac{1}{t}\right).$$

iii) Let Θ be a set of stationary and ergodic processes such that there exists a measure μ_Θ for which the estimation error of the probability goes to 0 uniformly:

$$\lim_{t\to\infty}\sup_{P\in\Theta}\left(\frac{1}{t}\sum_{u\in A^t} P(u)\log\left(P(u)/\mu_\Theta(u)\right)\right) = 0.$$

Then the error of the estimator which is based on the measure R goes to 0 uniformly too:

$$\lim_{t\to\infty}\sup_{P\in\Theta}\left(\frac{1}{t}\sum_{u\in A^t} P(u)\log(P(u)/R(u))\right) = 0.$$

1.3.2 Prediction

As we mentioned above, any universal code U can be applied for prediction. Namely, the measure μ_U (1.31) can be used for prediction as the following conditional probability:

$$\mu_U(x_{t+1}|x_1\dots x_t) = \mu_U(x_1\dots x_t x_{t+1})/\mu_U(x_1\dots x_t). \qquad (1.32)$$

The following theorem shows that such a predictor is quite reasonable. Moreover, it gives a possibility to apply practically used data compressors for prediction of real data and obtain quite precise estimation (see examples below).

Theorem 1.4 *Let U be a universal code and P be any stationary and ergodic process. Then*

$$i)\ \lim_{t\to\infty} \frac{1}{t} E\{\log \frac{P(x_1)}{\mu_U(x_1)} + \log \frac{P(x_2|x_1)}{\mu_U(x_2|x_1)} + \ldots + \log \frac{P(x_t|x_1 \ldots x_{t-1})}{\mu_U(x_t|x_1 \ldots x_{t-1})}\} = 0,$$

$$ii)\ \lim_{t\to\infty} E(\frac{1}{t} \sum_{i=0}^{t-1} (P(x_{i+1}|x_1 \ldots x_i) - \mu_U(x_{i+1}|x_1 \ldots x_i))^2) = 0,$$

and

$$iii)\ \lim_{t\to\infty} E(\frac{1}{t} \sum_{i=0}^{t-1} |P(x_{i+1}|x_1 \ldots x_i) - \mu_U(x_{i+1}|x_1 \ldots x_i)|) = 0.$$

An informal outline of the proof is as follows:

$$\frac{1}{t} \{E(\log \frac{P(x_1)}{\mu_U(x_1)}) + E(\log \frac{P(x_2|x_1)}{\mu_U(x_2|x_1)}) + \ldots + E(\log \frac{P(x_t|x_1 \ldots x_{t-1})}{\mu_U(x_t|x_1 \ldots x_{t-1})})\}$$

is equal to $\frac{1}{t} E(\log \frac{P(x_1 \ldots x_t)}{\mu_U(x_1 \ldots x_t)})$. Taking into account Theorem 1.2, we obtain the first statement of the theorem.

Comment 1.1 The measure R described above has one additional property if it is used for prediction. Namely, for any Markov process P ($P \in M^*(A)$) the following is true:

$$\lim_{t\to\infty} \log \frac{P(x_{t+1}|x_1 \ldots x_t)}{R(x_{t+1}|x_1 \ldots x_t)} = 0$$

with probability 1, where $R(x_{t+1}|x_1 \ldots x_t) = R(x_1 \ldots x_t x_{t+1})/R(x_1 \ldots x_t)$ [34].

Comment 1.2 It is known [44] that, in fact, the statements ii) and iii) are equivalent.

1.3.3 Problems with Side Information

Now we consider the so-called problems with side information, which are described as follows: there is a stationary and ergodic source whose alphabet A is presented as a product $A = X \times Y$. We are given a sequence $(x_1, y_1), \ldots, (x_{t-1}, y_{t-1})$ and side information y_t. The goal is to predict, or estimate, x_t. This problem arises in statistical decision theory, pattern recognition, and machine learning. Obviously, if someone knows the conditional probabilities $P(x_t|\ (x_1, y_1), \ldots, (x_{t-1}, y_{t-1}), y_t)$ for all $x_t \in X$, he has all information about x_t, available before x_t is known. That is why we will look for the best (or, at least, good) estimations for these conditional probabilities. Our solution will be based on results obtained in the previous subsection. More precisely, for any universal code U and the corresponding measure μ_U (1.31) we define the following estimate for the problem with side information:

$$\mu_U(x_t|(x_1, y_1), \ldots, (x_{t-1}, y_{t-1}), y_t) = \frac{\mu_U((x_1, y_1), \ldots, (x_{t-1}, y_{t-1}), (x_t, y_t))}{\sum_{x_t \in X} \mu_U((x_1, y_1), \ldots, (x_{t-1}, y_{t-1}), (x_t, y_t))}.$$

The following theorem shows that this estimate is quite reasonable.

Theorem 1.5 *Let U be a universal code and let P be any stationary and ergodic process. Then*

$$i)\ \lim_{t\to\infty} \frac{1}{t} \{E(\log \frac{P(x_1|y_1)}{\mu_U(x_1|y_1)}) + E(\log \frac{P(x_2|(x_1, y_1), y_2)}{\mu_U(x_2|(x_1, y_1), y_2)}) + \ldots$$
$$+E(\log \frac{P(x_t|(x_1, y_1), \ldots, (x_{t-1}, y_{t-1}), y_t)}{\mu_U(x_t|(x_1, y_1), \ldots, (x_{t-1}, y_{t-1}), y_t)})\} = 0,$$

$$ii)\ \lim_{t\to\infty} E(\frac{1}{t} \sum_{i=0}^{t-1} (P(x_{i+1}|(x_1, y_1), \ldots, (x_i, y_i), y_{i+1})-$$
$$\mu_U(x_{i+1}|(x_1, y_1), \ldots, (x_i, y_i), y_{i+1}))^2) = 0,$$

and

$$iii)\ \lim_{t\to\infty} E(\frac{1}{t} \sum_{i=0}^{t-1} |P(x_{i+1}|(x_1, y_1), \ldots, (x_i, y_i), y_{i+1})-$$
$$\mu_U(x_{i+1}|(x_1, y_1), \ldots, (x_i, y_i), y_{i+1})|) = 0.$$

The proof is very close to the proof of the previous theorem.

1.3.4 The Case of Several Independent Samples

In this section we consider a situation which is important for practical applications, but needs cumbersome notations. Namely, we extend our consideration to the case where the sample is presented as several independent samples $x^1 = x^1_1 \ldots x^1_{t_1}$, $x^2 = x^2_1 \ldots x^2_{t_2}, \ldots, x^r = x^r_1 \ldots x^r_{t_r}$ generated by a source. More precisely, we will suppose that all sequences were independently created by one stationary and ergodic source. (The point is that it is impossible just to combine all samples into one if the source is not i.i.d.) We denote them by $x^1 \diamond x^2 \diamond \ldots \diamond x^r$ and define $\nu_{x^1 \diamond x^2 \diamond \ldots \diamond x^r}(v) = \sum_{i=1}^r \nu_{x^i}(v)$. For example, if $x^1 = 0010, x^2 = 011$, then $\nu_{x^1 \diamond x^2}(00) = 1$. The definition of K_m and R can be extended to this case:

$$K_m(x^1 \diamond x^2 \diamond \ldots \diamond x^r) = \tag{1.33}$$

$$(\prod_{i=1}^{r} |A|^{-\min\{m,t_i\}}) \prod_{v \in A^m} \frac{\prod_{a \in A}((\Gamma(\nu_{x^1 \diamond x^2 \diamond \ldots \diamond x^r}(va) + 1/2) \,/\, \Gamma(1/2))}{(\Gamma(\bar{\nu}_{x^1 \diamond x^2 \diamond \ldots \diamond x^r}(v) + |A|/2) \,/\, \Gamma(|A|/2))},$$

whereas the definition of R is the same (see (1.25)). (Here, as before, $\bar{\nu}_{x^1 \diamond x^2 \diamond \ldots \diamond x^r}(v) = \sum_{a \in A} \nu_{x^1 \diamond x^2 \diamond \ldots \diamond x^r}(va)$. Note that $\bar{\nu}_{x^1 \diamond x^2 \diamond \ldots \diamond x^r}() = \sum_{i=1}^r t_i$ if $m = 0$.)

The following example is intended to show the difference between the cases of many samples and one.

Example Let there be two independent samples $y = y_1 \ldots y_4 = 0101$ and $x = x_1 \ldots x_3 = 101$, generated by a stationary and ergodic source with the alphabet $\{0, 1\}$. One wants to estimate the (limiting) probabilities $P(z_1 z_2), z_1, z_2 \in \{0, 1\}$ (here $z_1 z_2 \ldots$ can be considered as an independent sequence, generated by the source) and predict $x_4 x_5$ (i.e., estimate conditional probability $P(x_4 x_5 | x_1 \ldots x_3 = 101, y_1 \ldots y_4 = 0101)$. For solving both problems we will use the measure R (see (1.25)). First we consider the case where $P(z_1 z_2)$ is to be estimated without knowledge of sequences x and y. Those probabilities were calculated previously and we obtained: $R(00) \approx 0.296$, $R(01) = R(10) \approx 0.204$, $R(11) \approx 0.296$. Let us now estimate the probability $P(z_1 z_2)$ taking into account that there are two independent samples, $y = y_1 \ldots y_4 = 0101$ and $x = x_1 \ldots x_3 = 101$. First of all we note that such estimates are based on the formula for conditional probabilities:

$$R(z|x \diamond y) = R(x \diamond y \diamond z)/R(x \diamond y).$$

Then we estimate the frequencies: $\nu_{0101 \diamond 101}(0) = 3$, $\nu_{0101 \diamond 101}(1) = 4$, $\nu_{0101 \diamond 101}(00) = \nu_{0101 \diamond 101}(11) = 0$, $\nu_{0101 \diamond 101}(01) = 3$, $\nu_{0101 \diamond 101}(10) = 2$, $\nu_{0101 \diamond 101}(010) = 1$, $\nu_{0101 \diamond 101}(101) = 2$, $\nu_{0101 \diamond 101}(0101) = 1$, whereas frequencies of all other three-letter and four-letter words are 0. Then we calculate:

$$K_0(\,0101 \diamond 101) = \frac{1}{2}\frac{3}{4}\frac{5}{6}\frac{7}{8}\frac{1}{10}\frac{3}{12}\frac{5}{14} \approx 0.00244,$$

$$K_1(\,0101 \diamond 101) = (2^{-1})^2 \,\frac{1}{2}\frac{3}{4}\frac{5}{6}\; 1 \;\frac{1}{2}\frac{3}{4}\; 1$$

$$\approx 0.0293, \quad K_2(\,0101 \diamond 101) \approx 0.01172, \quad K_i(\,0101 \diamond 101) = 2^{-7},\ i \geq 3,$$

$$R(\,0101 \diamond 101) = \omega_1 K_0(\,0101 \diamond 101) + \omega_2 K_1(\,0101 \diamond 101) + \ldots \approx$$

$$0.369\, 0.00244 \;+\; 0.131\, 0.0293 \;+ 0.06932\, 0.01172 \;+ 2^{-7} \,/\, \log 5 \approx 0.0089.$$

In order to avoid repetitions, we estimate only one probability $P(z_1z_2 = 01)$. Carrying out similar calculations, we obtain $R(0101 \diamond 101 \diamond 01) \approx 0.00292$, $R(z_1z_2 = 01|y_1\dots y_4 = 0101, x_1\dots x_3 = 101) = R(0101 \diamond 101 \diamond 01)/R(0101 \diamond 101) \approx 0.32812$. If we compare this value and the estimation $R(01) \approx 0.204$, which is not based on the knowledge of samples x and y, we can see that the measure R uses additional information quite naturally (indeed, 01 is quite frequent in $y = y_1\dots y_4 = 0101$ and $x = x_1\dots x_3 = 101$).

Such generalization can be applied to many universal codes, but, generally speaking, there exist codes U for which $U(x^1 \diamond x^2)$ is not defined and, hence, the measure $\mu_U(x_1 \diamond x_2)$ is not defined. That is why we will describe properties of the universal code R, but not of universal codes in general. For the measure R all asymptotic properties are the same for the cases of one sample and several samples. More precisely, the following statement is true:

Claim 1.5 Let $x^1, x^2, \dots, x^r$ be independent sequences generated by a stationary and ergodic source and let t be a total length of these sequences ($t = \sum_{i=1}^r |x^i|$). Then, if $t \to \infty$ (and r is fixed) the statements of the Theorems 1.2–1.5 are valid when applied to $x^1 \diamond x^2 \diamond \dots \diamond x^r$ instead of $x_1\dots x_t$. (In Theorems 1.2–1.5 μ_U should be changed to R.)

The proofs are completely analogous to the proofs of the Theorems 1.2–1.5.

Now we can extend the definition of the empirical Shannon entropy (1.23) to the case of several words, $x^1 = x^1_1\dots x^1_{t_1}$, $x^2 = x^2_1\dots x^2_{t_2}, \dots, x^r = x^r_1\dots x^r_{t_r}$. We define $\nu_{x^1\diamond x^2\diamond\dots\diamond x^r}(v) = \sum_{i=1}^r \nu_{x^i}(v)$. For example, if $x^1 = 0010, x^2 = 011$, then $\nu_{x^1\diamond x^2}(00) = 1$. Analogously to (1.23),

$$h^*_k(x^1 \diamond x^2 \diamond \dots \diamond x^r) = -\sum_{v\in A^k} \frac{\bar\nu_{x^1\diamond\dots\diamond x^r}(v)}{(t-kr)} \sum_{a\in A} \frac{\nu_{x^1\diamond\dots\diamond x^r}(va)}{\bar\nu_{x^1\diamond\dots\diamond x^r}(v)} \log \frac{\nu_{x^1\diamond\dots\diamond x^r}(va)}{\bar\nu_{x^1\diamond\dots\diamond x^r}(v)}, \tag{1.34}$$

where $\bar\nu_{x^1\diamond\dots\diamond x^r}(v) = \sum_{a\in A} \nu_{x^1\diamond\dots\diamond x^r}(va)$.

For any sequence of words $x^1 = x^1_1\dots x^1_{t_1}$, $x^2 = x^2_1\dots x^2_{t_2}, \dots, x^r = x^r_1\dots x^r_{t_r}$ from A^* and any measure θ we define $\theta(x^1 \diamond x^2 \diamond \dots \diamond x^r) = \prod_{i=1}^r \theta(x^i)$. The following lemma gives an upper bound for unknown probabilities.

Lemma 1.1 *Let θ be a measure from $M_m(A), m \geq 0$, and $x^1, \dots, x^r$ be words from A^* whose lengths are not less than m. Then*

$$\theta(x^1 \diamond \dots \diamond x^r) \leq 2^{-(t-rm)\, h^*_m(x^1\diamond\dots\diamond x^t)}, \tag{1.35}$$

where $\theta(x^1 \diamond \dots \diamond x^r) = \prod_{i=1}^r \theta(x^i)$.

1.4 Hypothesis Testing

1.4.1 Goodness-of-Fit or Identity Testing

Now we consider the problem of testing H_0^{id} against H_1^{id}. Let us recall that the hypothesis H_0^{id} is that the source has a particular distribution π and the alternative hypothesis H_1^{id} that the sequence is generated by a stationary and ergodic source which differs from the source under H_0^{id}. Let the required level of significance (or the Type I error) be α, $\alpha \in (0, 1)$. We describe a statistical test which can be constructed based on any code φ.

The main idea of the suggested test is quite natural: compress a sample sequence $x_1 \dots x_t$ by a code φ. If the length of the codeword ($|\varphi(x_1 \dots x_t)|$) is significantly less than the value $-\log \pi(x_1 \dots x_t)$, then H_0^{id} should be rejected. The key observation is that the probability of all rejected sequences is quite small for any φ; that is why the Type I error can be made small. The precise description of the test is as follows: *The hypothesis H_0^{id} is accepted if*

$$- \log \pi(x_1 \dots x_t) - |\varphi(x_1 \dots x_t)| \leq - \log \alpha. \tag{1.36}$$

Otherwise, H_0^{id} is rejected. We denote this test by $T_\varphi^{id}(A, \alpha)$.

Theorem 1.6 *i) For each distribution $\pi, \alpha \in (0, 1)$ and a code φ, the Type I error of the described test $T_\varphi^{id}(A, \alpha)$ is not larger than α and ii) if, in addition, π is a finite-order stationary and ergodic process over A^∞ (i.e., $\pi \in M^*(A)$) and φ is a universal code, then the Type II error of the test $T_\varphi^{id}(A, \alpha)$ goes to 0 when t tends to infinity.*

1.4.2 Testing for Serial Independence

Let us recall that the null hypothesis H_0^{SI} is that the source is Markovian of order not larger than m ($m \geq 0$), and the alternative hypothesis H_1^{SI} is that the sequence is generated by a stationary and ergodic source which differs from the source under H_0^{SI}. In particular, if $m = 0$, this is the problem of testing for independence of time series.

Let there be given a sample $x_1 \dots x_t$ generated by an (unknown) source π. The main hypothesis H_0^{SI} is that the source π is Markovian whose order is not greater than m ($m \geq 0$), and the alternative hypothesis H_1^{SI} is that the sequence is generated by a stationary and ergodic source which differs from the source under H_0^{SI}. The described test is as follows.

Let φ be any code. By definition, the hypothesis H_0^{SI} is accepted if

$$(t-m)\, h^*_m(x_1 \ldots x_t) - |\varphi(x_1 \ldots x_t)| \le \log(1/\alpha)\,, \tag{1.37}$$

where $\alpha \in (0,1)$. Otherwise, H_0^{SI} is rejected. We denote this test by $T_\varphi^{SI}(A,\alpha)$.

Theorem 1.7 *i) For any code φ the Type I error of the test $T_\varphi^{SI}(A,\alpha)$ is less than or equal to $\alpha, \alpha \in (0,1)$ and ii) if, in addition, φ is a universal code, then the Type II error of the test $T_\varphi^{SI}(A,\alpha)$ goes to 0 when t tends to infinity.*

1.5 Examples of Hypothesis Testing

The goodness-of-fit testing will be considered in the next chapter, where the problem of randomness testing will be deeply investigated. In this part we consider a selection of the simulation results concerning independence tests.

We generated binary sequences by the first order Markov source with different probabilities (see Table 1.1) and applied the test $T_\varphi^{SI}(\{0,1\},\alpha)$ to test the hypothesis H_0^{SI} that a given bit sequence is generated by a Bernoulli source and the alternative hypothesis H_1^{SI} that the sequence is generated by a stationary and ergodic source which differs from the source under H_0^{SI}.

We tried several different archivers and the universal code R. It turned out that the power of the code R is larger than the power of the tried archivers; that is why we present results for the test $T_R^{SI}(\{0,1\},\alpha)$, which is based on this code, for $\alpha = 0.01$. Table 1.1 contains results of calculations.

We know that the source is Markovian and, hence, the hypothesis H_0^{SI} (that a sequence is generated by Bernoulli source) is not true. The table shows how the value of the Type II error depends on the sample size and the source probabilities.

Similar calculations were carried out for the Markov source of order 6. We applied the test $T_\varphi^{SI}(\{0,1\},\alpha)$, $\alpha = 0.01$, for checking the hypothesis H_0^{SI} that a given bit sequence is generated by Markov source of order at most 5 and the alternative hypothesis H_1^{SI} that the sequence is generated by a stationary and ergodic source which differs from the source under H_0^{SI}. Again, we know that H_0^{SI} is not true and Table 1.2 shows how the value of the Type II error depends on the sample size and the source probabilities.

Table 1.1 Serial independence testing for Markov source ("rej" means rejected, "acc" accepted. In all cases $p(x_{i+1} = 0|x_i = 1) = 0.5$)

Probabilities/length (bits)	2^9	2^{14}	2^{16}	2^{18}	2^{23}
$p(x_{i+1} = 0\|x_i = 0) = 0.8$	rej	rej	rej	rej	rej
$p(x_{i+1} = 0\|x_i = 0) = 0.6$	acc	rej	rej	rej	rej
$p(x_{i+1} = 0\|x_i = 0) = 0.55$	acc	acc	rej	rej	rej
$p(x_{i+1} = 0\|x_i = 0) = 0.525$	acc	acc	acc	rej	rej
$p(x_{i+1} = 0\|x_i = 0) = 0.505$	acc	acc	acc	acc	rej

Table 1.2 Serial independence testing for Markov source of order 6

Probabilities/length (bits)	2^{14}	2^{18}	2^{20}	2^{23}	2^{28}
$p(x_{i+1} = 0 \mid (\sum_{j=i-6}^{i} x_j) \, mod \, 2 = 0) = 0.8$	rej	rej	rej	rej	rej
$p(x_{i+1} = 0 \mid (\sum_{j=i-6}^{i} x_j) \, mod \, 2 = 0) = 0.6$	acc	rej	rej	rej	rej
$p(x_{i+1} = 0 \mid (\sum_{j=i-6}^{i} x_j) \, mod \, 2 = 0) = 0.55$	acc	acc	rej	rej	rej
$p(x_{i+1} = 0 \mid (\sum_{j=i-6}^{i} x_j) \, mod \, 2 = 0) = 0.525$	acc	acc	acc	rej	rej
$p(x_{i+1} = 0 \mid (\sum_{j=i-6}^{i} x_j) \, mod \, 2 = 0) = 0.505$	acc	acc	acc	acc	rej

In all cases $p(x_{i+1} = 0 \mid (\sum_{j=i-6}^{i} x_i) \, mod \, 2 = 1) = 0.5$

1.6 Real-Valued Time Series

1.6.1 Density Estimation and Its Application

Here we address the problem of nonparametric estimation of the density for time series. Let X_t be a time series and the probability distribution of X_t be unknown; but it is known that the time series is stationary and ergodic. We have seen that the Shannon-McMillan-Breiman theorem played a key role in the case of finite-alphabet processes. In this part we will use its generalization to the processes with densities, which was established by Barron [3]. First we describe considered processes with some properties needed for the generalized Shannon-MacMillan-Breiman theorem to hold. In what follows, we restrict our attention to processes that take bounded real values. However, the main results may be extended to processes taking values in a compact subset of a separable metric space.

Let B denote the Borel subsets of R, and B^k denote the Borel subsets of R^k, where R is the set of real numbers. Let R^∞ be the set of all infinite sequences $x = x_1, x_2 \dots$ with $x_i \in \mathsf{R}$, and let B^∞ denote the usual product sigma field on R^∞, generated by the finite dimensional cylinder sets $\{A_1, \dots A_k, \mathsf{R}, \mathsf{R}, \dots\}$, where $A_i \in B, i = 1, \dots, k$. Each stochastic process $X_1, X_2, \dots, X_i \in \mathsf{R}$, is defined by a probability distribution on $(\mathsf{R}^\infty, B^\infty)$. Suppose that the joint distribution P_n for $(X_1, X_2, \dots, X_n)$ has a probability density function $p(x_1 x_2 \dots x_n)$ with respect to a sigma-finite measure M_n. Assume that the sequence of dominating measures M_n is Markov of order $m \geq 0$ with a stationary transition measure. (For example, M_n can be the Lebesgue measure, counting measure, etc.) Let $p(x_{n+1}|x_1 \dots x_n)$ denote the conditional density given by the ratio $p(x_1 \dots x_{n+1}) \, / p(x_1 \dots x_n)$ for $n > 1$. It is known that for stationary and ergodic processes there exists a so-called relative entropy rate $\tilde{h}$ defined by

$$\tilde{h} = \lim_{n \to \infty} -E(\log p(x_{n+1}|x_1 \dots x_n)), \tag{1.38}$$

where E denotes expectation with respect to P. We will use the following generalization of the Shannon-MacMillan-Breiman theorem:

Claim 1.6 ([3]) If $\{X_n\}$ is a P-stationary ergodic process with density $p(x_1 \dots x_n) = dP_n/dM_n$ and $\tilde{h}_n < \infty$ for some $n \geq m$, the sequence of relative entropy densities $-(1/n)\log p(x_1 \dots x_n)$ converges almost surely to the relative entropy rate, i.e.,

$$\lim_{n\to\infty} (-1/n) \log p(x_1 \dots x_n) = \tilde{h} \tag{1.39}$$

with probability 1 (according to P).

Now we return to the estimation problems. Let $\{\Pi_n\}, n \geq 1$, be an increasing sequence of finite partitions of R that asymptotically generates the Borel sigma-field B and let $x^{[k]}$ denote the element of Π_k that contains the point x. (Informally, $x^{[k]}$ is obtained by quantizing x to k bits of precision.) For integers s and n we define the following approximation of the density:

$$p^s(x_1 \dots x_n) = P(x_1^{[s]} \dots x_n^{[s]})/M_n(x_1^{[s]} \dots x_n^{[s]}). \tag{1.40}$$

We also consider

$$\tilde{h}_s = \lim_{n\to\infty} -E(\log p^s(x_{n+1}|x_1 \dots x_n)). \tag{1.41}$$

Applying the claim 1.6 to the density $p^s(x_1 \dots x_t)$, we obtain that a.s.

$$\lim_{t\to\infty} -\frac{1}{t} \log p^s(x_1 \dots x_t) = \tilde{h}_s. \tag{1.42}$$

Let U be a universal code, which is defined for any finite alphabet. In order to describe a density estimate we will use the probability distribution $\omega_i, i = 1, 2, \dots;$ see (1.24). (In what follows we will use this distribution, but results described below are obviously true for any distribution with nonzero probabilities.) Now we can define the density estimate r_U as follows:

$$r_U(x_1 \dots x_t) = \sum_{i=0}^{\infty} \omega_i \, \mu_U(x_1^{[i]} \dots x_t^{[i]})/M_t(x_1^{[i]} \dots x_t^{[i]}) \,, \tag{1.43}$$

where the measure μ_U is defined by (1.31). (It is assumed here that the code $U(x_1^{[i]} \dots x_t^{[i]})$ is defined for the alphabet, which contains $|\Pi_i|$ letters.)

It turns out that, in a certain sense, the density $r_U(x_1 \dots x_t)$ estimates the unknown density $p(x_1 \dots x_t)$.

Theorem 1.8 *Let X_t be a stationary ergodic process with densities $p(x_1 \dots x_t) = dP_t/dM_t$ such that*

$$\lim_{s\to\infty} \tilde{h}_s = \tilde{h} < \infty, \tag{1.44}$$

where $\tilde{h}$ and $\tilde{h}_s$ are relative entropy rates; see (1.38), (1.41). Then

$$\lim_{t\to\infty} \frac{1}{t} \log \frac{p(x_1 \dots x_t)}{r_U(x_1 \dots x_t)} = 0 \tag{1.45}$$

with probability 1 and

$$\lim_{t\to\infty} \frac{1}{t}\, E(\log \frac{p(x_1 \dots x_t)}{r_U(x_1 \dots x_t)}) = 0\,. \tag{1.46}$$

We have seen that the requirement (1.44) plays an important role in the proof. The natural question is whether there exist processes for which (1.44) is valid. The answer is positive. For example, let a process possess values in the interval $[-1, 1]$, M_n be Lebesgue measure and the considered process be Markovian with conditional density

$$p(x|y) = \begin{cases} 1/2 + \alpha \;\; sign(y), & \text{if } x < 0, \\ 1/2 - \alpha \;\; sign(y), & \text{if } x \geq 0, \end{cases}$$

where $\alpha \in (0, 1/2)$ is a parameter and

$$sign(y) = \begin{cases} -1, & \text{if } y < 0, \\ 1, & \text{if } y \geq 0. \end{cases}$$

In words, the density depends on a sign of the previous value. If the value is positive, then the density is more than 1/2, otherwise it is less than 1/2. It is easy to see that (1.44) is true for any $\alpha \in (0, 1)$.

The following two theorems are devoted to the conditional probability $r_U(x|x_1 \dots x_m) = r_U(x_1 \dots x_m x)/\, r_U(x_1 \dots x_m)$ which, in turn, is connected with the prediction problem. We will see that the conditional density $r_U(x|x_1 \dots x_m)$ is a reasonable estimation of the unknown density $p(x|x_1 \dots x_m)$.

Theorem 1.9 *Let $B_1, B_2, \ldots$ be a sequence of measurable sets. Then the following equalities are true:*

$$i)\ \lim_{t\to\infty} E\Big(\frac{1}{t}\sum_{m=0}^{t-1}(P(x_{m+1} \in B_{m+1}|x_1 \ldots x_m) - R_U(x_{m+1} \in B_{m+1}|x_1 \ldots x_m))^2\Big) = 0\,, \tag{1.47}$$

$$ii)\ E\Big(\frac{1}{t}\sum_{m=0}^{t-1}|P(x_{m+1} \in B_{m+1}|x_1 \ldots x_m) - R_U(x_{m+1} \in B_{m+1}|x_1 \ldots x_m))|\Big) = 0\,,$$

where $R_U(x_{m+1} \in B_{m+1}|x_1 \ldots x_m) = \int_{B_{m+1}} r_U(x|x_1 \ldots x_m)dM_{1/m}$.

We have seen that in a certain sense the estimation r_U approximates the unknown density p. The following theorem shows that r_U can be used instead of p for estimation of average values of certain functions.

Theorem 1.10 *Let f be an integrable function whose absolute value is bounded by a certain constant $\bar{M}$ and all conditions of Theorem 1.2 are true. Then the following equality is valid:*

$$i)\ \lim_{t\to\infty}\frac{1}{t}E\Big(\sum_{m=0}^{t-1}\Big(\int f(x)\,p(x|x_1\ldots x_m)dM_m - \int f(x)\,r_U(x|x_1\ldots x_m)dM_m\Big)^2\Big) = 0, \tag{1.48}$$

$$ii)\ \lim_{t\to\infty}\frac{1}{t}E\Big(\sum_{m=0}^{t-1}\Big|\int f(x)\,p(x|x_1\ldots x_m)\,dM_m - \int f(x)\,r_U(x|x_1\ldots x_m)\,dM_m\Big|\Big) = 0.$$

1.6.2 Example of Forecasting

We consider time series generating numbers from an interval $[A, B]$. Let $\{\Pi_n\}$, $n \geq 1$ be an increasing sequence of finite partitions of the interval $[A, B]$ such that the maximum size of a subinterval in the partitions goes to zero when m grows. Denote the element of the partition Π_k containing the point x by $x^{[k]}$.

In the following we will only consider processes for which all multidimensional densities with respect to Lebesgue measure L exist. From (1.43),

$$r\,(x_1 \ldots x_t) = \sum_{s=1}^{\infty} \omega_s \mathrm{R}(x_1{}^{[s]} \ldots x_t{}^{[s]})/\mathrm{L}(x_1{}^{[s]} \ldots x_t{}^{[s]}), \tag{1.49}$$

where ω_s are defined by (1.24) and s is the number of the partition Π_s.

1.6.2.1 The Algorithm and Its Complexity

It is possible to use any sequence of finite partitions, but experimentally we found out that partitioning into equal subintervals gives the best accuracy for the forecast on real data; that is why in the experiments described below such partitions were used. Denote the partitioning of the interval $[A, B]$ into n equal subintervals as Π_n.

We will estimate a complexity of algorithms by the number of operations required for calculating the conditional density

$$r(a|x_1 \dots x_t) = r(x_1 \dots x_t a)/r(x_1 \dots x_t). \tag{1.50}$$

For the partitioning Π_n this number of operations is determined by the complexity of calculation of the measure R in an n-letter alphabet, which, in turn, depends on the complexity of calculating K_m; see (1.19). Having taken into account that $K_m = n^{-t}$ if $m > t$, we can see that the complexity of calculation of K_m, $m = 1, 2, \dots$ in (1.19) is equal to $O(t\, n^{t+1})$, $t \to \infty$, where n is the number of subintervals in the partition, and t is the length of the row $x_1 \dots x_t$. From this we obtain that the complexity of estimation of the conditional density (1.50) is equal to $O(t^3\, n^{t+2})$, $t \to \infty$. So, we can see that the number of the subintervals of the partition (n) determines the complexity of the algorithm. It turns out that the complexity can be reduced if n is large. The fact that in this case many frequencies of occurrence of subintervals (ν in (1.19)) coincide allows us to use the method of grouping of alphabet letters from [38]. In this case, the reduction of complexity cannot be described analytically since this value, generally speaking, depends on the considered time series. However, experiments showed that the time of calculations was reduced by a factor of 3–5, with the length of a row from several hundred to two thousand elements. So, on the one hand, the number of subintervals of the partition (n) determines the precision of the forecast, but, on the other hand, the complexity of the algorithm depends sufficiently on this n. Experimentally we determined that the value of $n = \lfloor \log_2 t \rfloor + 5$ is a good choice, because larger values of n almost do not change the precision, but considerably increase the time of calculation. So, in our calculations we used the estimator

$$r(x_1 \dots x_t) = \sum_{s=1}^{\lfloor \log_2 t \rfloor + 4} \omega_s R({x_1}^{[s]} \dots {x_t}^{[s]})/\mathrm{L}({x_1}^{[s]} \dots {x_t}^{[s]}) +$$

$$(1/(\lfloor \log_2 t \rfloor + 6)) \; R({x_1}^{[\lfloor \log_2 t \rfloor + 5]} \dots {x_t}^{[\lfloor \log_2 t \rfloor + 5]})/\mathrm{L}({x_1}^{[\lfloor \log_2 t \rfloor + 5]} \dots {x_t}^{[\lfloor \log_2 t \rfloor + 5]}) \tag{1.51}$$

instead of (1.43).

1.6.2.2 The Experiments

We used the following scheme of experiments: for a data sequence $x_1 \ldots x_t$ we took $t - 10$ first points $x_1 \ldots x_{t-10}$ and calculated the density estimation (1.50) for $p(x_{t-9}/x_1 \ldots x_{t-10})$. Then, based on this density, we calculated the mean value (denoted y_{t-9}) and the difference $d_1 = |x_{t-9} - y_{t-9}|$. Analogously we calculated y_{t-8} and d_2 based on $x_1 \ldots x_{t-9}$, etc. Then we calculated the following mean forecasting error $\hat{d} = \sum_{i=1}^{10} |x_i - y_i|)/10$.

As an example we consider forecasting of the production price index in the US [5]; see Fig. 1.1. The observation period is from 01.1990 to 02.2013; the interval is one month, and the length of the row (t) equals 277. It turns out that the mean forecasting error $\hat{d}$ equals 0.912, whereas the maximal difference between two neighboring values equals 8.87 (i.e., $\max_{i=2,\ldots,t} |x_i - x_{i-1}| = 8.87$). So, we can see that forecast error is about 10 % of this value.

In order to compare the suggested method and known ones we took data on economic time series of the US, for which the results of forecasting methods of the International Institute of Forecasters (IIF) are available. We took the three following time series from [5]: finance (1) and finance (2) (index of financial activity in the US) and demographic (demographic indicators of the US). We calculated the mean error of the forecasting $\hat{d}$ described above, and compared it to the forecasting methods AutoBox, ForecastPro, PP-Autocast; see IIF [16]. The results of this comparison are presented in Table 1.3.

From this table we can see that the prediction error of the proposed method is significantly lower than that of the comparison methods.

Fig. 1.1 Graph of production price index in USA (interval: 1 month)

Table 1.3 Comparison of the proposed method with Autobox, ForecastPro and PP-Autocast

Time series	Length	Proposed method	Autobox	ForecastPro	PP-Autocast
Finance (1)	144	164.48	680.49	794.42	793.03
Finance (2)	132	21.07	76.12	71.98	41.40
Demographic	134	53.46	122.08	152.71	286.19

1.7 The Hypothesis Testing for Infinite Alphabet

In this subsection we consider a case where the source alphabet A is infinite, say, a part of R^n. Our strategy is to use finite partitions of A and to consider hypotheses corresponding to the partitions. This approach can be directly applied to the goodness-of-fit testing, but it cannot be applied to the serial independence testing. The point is that if someone combines letters (or states) of a Markov chain, the chain order (or memory) can increase. For example, if the alphabet contains three letters, there exists a Markov chain of order 1, such that combining two letters into one transforms the chain into a process with infinite memory. That is why in this part we will consider the independence testing for i.i.d. processes only (i.e., processes from $M_0(A)$).

In order to avoid repetitions, we will consider a general scheme, which can be applied to both tests using notations $H_0^{\aleph}, H_1^{\aleph}$ and $T_{\varphi}^{\aleph}(A, \alpha)$, where $\aleph$ is an abbreviation of one of the described tests (i.e., *id* and *SI*.)

Let us give some definitions. Let $\Lambda = \lambda_1, \ldots, \lambda_s$ be a finite (measurable) partition of A and let $\Lambda(x)$ be an element of the partition Λ which contains $x \in A$. For any process π we define a process π_Λ over a new alphabet Λ by the equation

$$\pi_\Lambda(\lambda_{i_1} \ldots \lambda_{i_k}) = \pi(x_1 \in \lambda_{i_1}, \ldots, x_k \in \lambda_{i_k}),$$

where $x_1 \ldots x_k \in A^k$.

We will consider an infinite sequence of partitions $\hat{\Lambda} = \Lambda_1, \Lambda_2, \ldots$. and say that such a sequence discriminates between a pair of hypotheses $H_0^{\aleph}(A), H_1^{\aleph}(A)$ about processes if for each process ϱ, for which $H_1^{\aleph}(A)$ is true there exists a partition Λ_j for which $H_1^{\aleph}(\Lambda_j)$ is true for the process ϱ_{Λ_j}.

Let $H_0^{\aleph}(A), H_1(A)^{\aleph}$ be a pair of hypotheses, $\hat{\Lambda} = \Lambda_1, \Lambda_2, \ldots$ be a sequence of partitions, α be from $(0, 1)$ and φ be a code. The scheme for both tests is as follows:

The hypothesis $H_0^{\aleph}(A)$ is accepted if for all $i = 1, 2, 3, \ldots$ the test $T_{\varphi}^{\aleph}(\Lambda_i, (\alpha\omega_i))$ accepts the hypothesis $H_0^{\aleph}(\Lambda_i)$. Otherwise, $H_0^{\aleph}$ is rejected. We denote this test by $\mathbf{T}^{\aleph}_{\alpha,\varphi}(\hat{\Lambda})$.

Comment 1.3 It is important to note that one does not need to check an infinite number of inequalities when applying this test. The point is that the hypothesis $H_0^{\aleph}(A)$ has to be accepted if the left part in (1.36) or (1.37) is less than $-\log(\alpha\omega_i)$. Obviously, $-\log(\alpha\omega_i)$ goes to infinity if i increases. That is why there are many cases, where it is enough to check a finite number of hypotheses $H_0^{\aleph}(\Lambda_i)$.

Theorem 1.11 *i) For each* $\alpha \in (0, 1)$, *sequence of partitions* $\hat{\Lambda}$ *and a code* φ, *the Type I error of the described test* $\boldsymbol{T}^{\aleph}_{\alpha,\varphi}(\hat{\Lambda})$ *is not larger than* α, *and ii) if, in addition,* φ *is a universal code and* $\hat{\Lambda}$ *discriminates between* $H_0^{\aleph}(A), H_1(A)^{\aleph}$, *then the Type II error of the test* $\boldsymbol{T}^{\aleph}_{\alpha,\varphi}(\hat{\Lambda})$ *goes to 0 when the sample size tends to infinity.*

1.8 Conclusion

Time series is a popular model of real stochastic processes which has many applications in industry, economy, meteorology and many other fields. Despite this, there are many practically important problems of statistical analysis of time series which are still open. Among them we can name the problems of estimation of the limiting probabilities and densities, on-line prediction, regression, and classification and some problems of hypothesis testing (goodness-of-fit testing and testing of serial independence). This chapter describes a new approach to all the problems mentioned above, which, on the one hand, gives a possibility to solve the problems in the framework of classical mathematical statistics and, on the other hand, allows us to apply methods of real data compression to solve these problems in practice.

Appendix

Proof of Claim 1.1 We employ the general inequality

$$D(\mu\|\eta) \leq \log e \ (-1 + \sum_{a \in A} \mu(a)^2/\eta(a)\,),$$

valid for any distributions μ and η over A (follows from the elementary inequality for natural logarithm $\ln x \leq x - 1$), and find:

$$\rho^t(P\|L_0) = \sum_{x_1 \cdots x_t \in A^t} P(x_1 \cdots x_t) \sum_{a \in A} P(a|x_1 \cdots x_t) \log \frac{P(a|x_1 \cdots x_t)}{\gamma(a|x_1 \cdots x_t)}$$

$$= \log e \,(\sum_{x_1 \cdots x_t \in A^t} P(x_1 \cdots x_t) \sum_{a \in A} P(a|x_1 \cdots x_t) \ln \frac{P(a|x_1 \cdots x_t)}{\gamma(a|x_1 \cdots x_t)})$$

$$\leq \log e \,(-1 + \sum_{x_1 \cdots x_t \in A^t} P(x_1 \cdots x_t) \sum_{a \in A} \frac{P(a)^2 (t + |A|)}{\nu_{x_1 \cdots x_t}(a) + 1}$$

Applying the well-known Bernoulli formula, we obtain

$$\rho^t(P\|L_0) = \log e\,(-1 + \sum_{a\in A}\sum_{i=0}^{t} \frac{P(a)^2(t+|A|)}{i+1} \binom{t}{i} P(a)^i(1-P(a))^{t-i})$$

$$= \log e\,(-1 + \frac{t+|A|}{t+1}\sum_{a\in A} P(a) \sum_{i=0}^{t} \binom{t+1}{i+1} P(a)^{i+1}(1-P(a))^{t-i})$$

$$\le \log e\,(-1 + \frac{t+|A|}{t+1}\sum_{a\in A} P(a) \sum_{j=0}^{t+1} \binom{t+1}{j} P(a)^{j}(1-P(a))^{t+1-j}).$$

Again, using the Bernoulli formula, we finish the proof:

$$\rho^t(P\|L_0) = \log e\,\frac{|A|-1}{t+1}.$$

The second statement of the claim follows from the well-known asymptotic equality

$$1 + 1/2 + 1/3 + \ldots + 1/t = \ln t + O(1),$$

the obvious presentation

$$\bar{\rho}^t(P\|L_0) = t^{-1}(\rho^0(P\|L_0) + \rho^1(P\|L_0) + \ldots + \rho^{t-1}(P\|L_0))$$

and (1.10).

Proof of Claim 1.2 The first equality follows from the definition (1.9), whereas the second from the definition (1.12). From (1.16) we obtain:

$$-\log K_0(x_1\ldots x_t) = -\log(\frac{\Gamma(|A|/2)}{\Gamma(1/2)^{|A|}}\,\frac{\prod_{a\in A}\Gamma(\nu^t(a)+1/2)}{\Gamma((t+|A|/2)})$$

$$= c_1 + c_2|A| + \log\Gamma(t+|A|/2) - \sum_{a\in A}\Gamma(\nu^t(a)+1/2),$$

where c_1, c_2 are constants. Now we use the well-known Stirling formula

$$\ln\Gamma(s) = \ln\sqrt{2\pi} + (s-1/2)\ln s - s + \theta/12,$$

where $\theta \in (0,1)$ [23] . Using this formula we rewrite the previous equality as

$$-\log K_0(x_1\ldots x_t) = -\sum_{a\in A}\nu^t(a)\log(\nu^t(a)/t) + (|A|-1)\log t/2 + \bar{c}_1 + \bar{c}_2|A|,$$

where $\bar{c}_1, \bar{c}_2$ are constants. Hence,

$$\sum_{x_1 \ldots x_t \in A^t} P(x_1 \ldots x_t)(-\log(K_0(x_1 \ldots x_t)))$$

$$\leq t(\sum_{x_1 \ldots x_t \in A^t} P(x_1 \ldots x_t)(-\sum_{a \in A} \nu^t(a) \log(\nu^t(a)/t)) + (|A| - 1) \log t/2 + c|A|.$$

Applying the well-known Jensen inequality for the concave function $-x \log x$ we obtain the following inequality:

$$\sum_{x_1 \ldots x_t \in A^t} P(x_1 \ldots x_t)(-\log(K_0(x_1 \ldots x_t)) \leq$$

$$-t(\sum_{x_1 \ldots x_t \in A^t} P(x_1 \ldots x_t)((\nu^t(a)/t))$$

$$\log \sum_{x_1 \ldots x_t \in A^t} P(x_1 \ldots x_t)(\nu^t(a)/t) + (|A| - 1) \log t/2 + c|A|.$$

The source P is i.i.d.; that is why the average frequency

$$\sum_{x_1 \ldots x_t \in A^t} P(x_1 \ldots x_t)\nu^t(a)$$

is equal to $P(a)$ for any $a \in A$ and we obtain from the last two formulas the following inequality:

$$\sum_{x_1 \ldots x_t \in A^t} P(x_1 \ldots x_t)(-\log(K_0(x_1 \ldots x_t))$$

$$\leq t(-\sum_{a \in A} P(a) \log P(a)) + (|A| - 1) \log t/2 + c|A|. \quad (1.52)$$

On the other hand,

$$\sum_{x_1 \ldots x_t \in A^t} P(x_1 \ldots x_t)(\log P(x_1 \ldots x_t)) = \sum_{x_1 \ldots x_t \in A^t} P(x_1 \ldots x_t) \sum_{i=1}^{t} \log P(x_i)$$

$$= t(\sum_{a \in A} P(a) \log P(a)). \quad (1.53)$$

From (1.52) and (1.53) we can see that

$$t^{-1} \sum_{x_1 \ldots x_t \in A^t} P(x_1 \ldots x_t) \log \frac{P(x_1 \ldots x_t)}{(K_0(x_1 \ldots x_t)} \leq ((|A| - 1) \log t/2 + c)/t.$$

Proof of Claim 1.3 First we consider the case where $m = 0$. The proof for this case is very close to the proof of the previous claim. Namely, from (1.16) we obtain:

$$-\log K_0(x_1 \dots x_t) = -\log(\frac{\Gamma(|A|/2)}{\Gamma(1/2)^{|A|}} \frac{\prod_{a\in A}\Gamma(\nu^t(a)+1/2)}{\Gamma((t+|A|/2)})$$

$$= c_1 + c_2|A| + \log \Gamma(t+|A|/2) - \sum_{a\in A} \Gamma(\nu^t(a)+1/2),$$

where c_1, c_2 are constants. Now we use the well-known Stirling formula

$$\ln \Gamma(s) = \ln \sqrt{2\pi} + (s - 1/2)\ln s - s + \theta/12,$$

where $\theta \in (0, 1)$ [23] . Using this formula we rewrite the previous equality as

$$-\log K_0(x_1 \dots x_t) = -\sum_{a\in A} \nu^t(a)\log(\nu^t(a)/t) + (|A|-1)\log t/2 + \bar{c}_1 + \bar{c}_2|A|,$$

where $\bar{c}_1, \bar{c}_2$ are constants. Having taken into account the definition of the empirical entropy (1.23), we obtain

$$-\log K_0(x_1 \dots x_t) \le th_0^*(x_1 \dots x_t) + (|A|-1)\log t/2 + c|A|.$$

Hence,

$$\sum_{x_1\dots x_t\in A^t} P(x_1 \dots x_t)(-\log(K_0(x_1 \dots x_t)))$$

$$\le t(\sum_{x_1\dots x_t\in A^t} P(x_1 \dots x_t)h_0^*(x_1 \dots x_t) + (|A|-1)\log t/2 + c|A|.$$

Having taken into account the definition of the empirical entropy (1.23), we apply the well-known Jensen inequality for the concave function $-x\log x$ and obtain the following inequality:

$$\sum_{x_1\dots x_t\in A^t} P(x_1 \dots x_t)(-\log(K_0(x_1 \dots x_t)) \le +c|A| -$$

$$t(\sum_{x_1\dots x_t\in A^t} P(x_1 \dots x_t)((\nu^t(a)/t))\log \sum_{x_1\dots x_t\in A^t} P(x_1 \dots x_t)(\nu^t(a)/t) + (|A|-1)\log t/2.$$

P is stationary and ergodic; that is why the average frequency

$$\sum_{x_1\dots x_t\in A^t} P(x_1 \dots x_t)\nu^t(a)$$

is equal to $P(a)$ for any $a \in A$ and we obtain from the last two formulas the following inequality:

$$\sum_{x_1 \ldots x_t \in A^t} P(x_1 \ldots x_t)(-\log(K_0(x_1 \ldots x_t)) \leq t\, h_0(P) + (|A| - 1) \log t/2 + c|A|,$$

where $h_0(P)$ is the first order Shannon entropy; see (1.12).

We have seen that any source from $M_m(A)$ can be presented as a "sum" of $|A|^m$ i.i.d. sources. From this we can easily see that the error of a predictor for the source from $M_m(A)$ can be upper bounded by the error of the i.i.d. source multiplied by $|A|^m$. In particular, we obtain from the last inequality and the definition of the Shannon entropy (1.20) the upper bound (1.22).

Proof (Theorem 1.1) We can see from the definition (1.25) of R and (1.19) that the average error is upper bounded as follows:

$$-t^{-1} \sum_{x_1 \ldots x_t \in A^t} P(x_1 \ldots x_t) \log(R(x_1 \ldots x_t)) - h_k(P)$$
$$\leq (|A|^k(|A| - 1) \log t + \log(1/\omega_i) + C)/(2t),$$

for any $k = 0, 1, 2, \ldots$. Taking into account that for any $P \in M_\infty(A)$ $\lim_{k\to\infty} h_k(P) = h_\infty(P)$, we can see that

$$(\lim_{t\to\infty} t^{-1} \sum_{x_1 \ldots x_t \in A^t} P(x_1 \ldots x_t) \log(R(x_1 \ldots x_t)) - h_\infty(P)) = 0.$$

The second statement of the theorem is proven. The first one can be easily derived from the ergodicity of P [4, 14] .

Proof (Theorem 1.2) The proof is based on the Shannon-MacMillan-Breiman theorem, which states that for any stationary and ergodic source P

$$\lim_{t\to\infty} -\log P(x_1 \ldots x_t)/t = h_\infty(P)$$

with probability 1 [4, 14]. From this equality and (1.29) we obtain the statement i). The second statement follows from the definition of the Shannon entropy (1.21) and (1.30).

Proof (Theorem 1.4) i) immediately follows from the second statement of Theorem 1.2 and properties of log. The statement ii) can be proven as follows:

$$\lim_{t\to\infty} E(\frac{1}{t}\sum_{i=0}^{t-1}(P(x_{i+1}|x_1\ldots x_i) - \mu_U(x_{i+1}|x_1\ldots x_i))^2) =$$

$$\lim_{t\to\infty}\frac{1}{t}\sum_{i=0}^{t-1}\sum_{x_1\ldots x_i\in A^i} P(x_1\ldots x_i)(\sum_{a\in A}|P(a|x_1\ldots x_i) - \mu_U(a|x_1\ldots x_i)|)^2 \le$$

$$\lim_{t\to\infty}\frac{const}{t}\sum_{i=0}^{t-1}\sum_{x_1\ldots x_i\in A^i} P(x_1\ldots x_i)\sum_{a\in A} P(a|x_1\ldots x_i)\log\frac{P(a|x_1\ldots x_i)}{\mu_U(a|x_1\ldots x_i)} =$$

$$\lim_{t\to\infty}(\frac{const}{t}\sum_{x_1\ldots x_t\in A^t} P(x_1\ldots x_t)\log(P(x_1\ldots x_t)/\mu(x_1\ldots x_t))).$$

Here the first inequality is obvious, the second follows from the Pinsker's inequality (1.5), the others from properties of expectation and log . iii) can be derived from ii) and the Jensen inequality for the function x^2.

Proof (Theorem 1.5) The following inequality follows from the nonnegativity of the KL divergency (see (1.5)), whereas the equality is obvious.

$$E(\log\frac{P(x_1|y_1)}{\mu_U(x_1|y_1)}) + E(\log\frac{P(x_2|(x_1,y_1),y_2)}{\mu_U(x_2|(x_1,y_1),y_2)}) + \ldots \le E(\log\frac{P(y_1)}{\mu_U(y_1)})$$

$$+E(\log\frac{P(x_1|y_1)}{\mu_U(x_1|y_1)}) + E(\log\frac{P(y_2|(x_1,y_1)}{\mu_U(y_2|(x_1,y_1)}) + E(\log\frac{P(x_2|(x_1,y_1),y_2)}{\mu_U(x_2|(x_1,y_1),y_2)}) + \ldots$$

$$= E(\log\frac{P(x_1,y_1)}{\mu_U(x_1,y_1)}) + E(\log\frac{P((x_2,y_2)|(x_1,y_1))}{\mu_U((x_2,y_2)|(x_1,y_1))}) + \ldots.$$

Now we can apply the first statement of the previous theorem to the last sum as follows:

$$\lim_{t\to\infty}\frac{1}{t}E(\log\frac{P(x_1,y_1)}{\mu_U(x_1,y_1)}) + E(\log\frac{P((x_2,y_2)|(x_1,y_1))}{\mu_U((x_2,y_2)|(x_1,y_1))}) + \ldots$$

$$E(\log\frac{P((x_t,y_t)|(x_1,y_1)\ldots(x_{t-1},y_{t-1}))}{\mu_U((x_t,y_t)|(x_1,y_1)\ldots(x_{t-1},y_{t-1}))}) = 0.$$

From this equality and the last inequality we obtain the proof of i). The proof of the second statement can be obtained from the similar representation for ii) and the second statement of Theorem 4. iii) can be derived from ii) and the Jensen inequality for the function x^2.

Proof (Lemma 1.1) First we show that for any source $\theta^* \in M_0(A)$ and any words $x^1 = x_1^1 \dots x_{t_1}^1, \dots, x^r = x_1^r \dots x_{t_r}^r$,

$$\theta^*(x^1 \diamond \dots \diamond x^r) = \prod_{a\in A} (\theta^*(a))^{\nu_{x^1\diamond\dots\diamond x^r}(a)}$$

$$\le \prod_{a\in A} (\nu_{x^1\diamond\dots\diamond x^r}(a)/t)^{\nu_{x^1\diamond\dots\diamond x^r}(a)}, \tag{1.54}$$

where $t = \sum_{i=1}^r t_i$. Here the equality holds, because $\theta^* \in M_0(A)$. The inequality follows from the definition of the KL divergence (1.4) if we take into account that the KL divergence is nonnegative. Indeed, if $p(a) = \nu_{x^1\diamond\dots\diamond x^r}(a)/t$ and $q(a) = \theta^*(a)$, then

$$\sum_{a\in A} \frac{\nu_{x^1\diamond\dots\diamond x^r}(a)}{t} \log \frac{(\nu_{x^1\diamond\dots\diamond x^r}(a)/t)}{\theta^*(a)} \ge 0.$$

From the latter inequality we obtain (1.54). Taking into account the definition (1.34) and (1.54), we can see that the statement of Lemma 1.1 is true for this particular case.

For any $\theta \in M_m(A)$ and $x = x_1 \dots x_s$, $s > m$, we present $\theta(x_1 \dots x_s)$ as

$$\theta(x_1 \dots x_s) = \theta(x_1 \dots x_m) \prod_{u\in A^m} \prod_{a\in A} \theta(a/u)^{\nu_x(ua)},$$

where $\theta(x_1 \dots x_m)$ is the limiting probability of the word $x_1 \dots x_m$. Hence, $\theta(x_1 \dots x_s) \le \prod_{u\in A^m} \prod_{a\in A} \theta(a/u)^{\nu_x(ua)}$. Taking into account the inequality (1.54), we obtain $\prod_{a\in A} \theta(a/u)^{\nu_x(ua)} \le \prod_{a\in A} (\nu_x(ua)/\bar{\nu}_x(u))^{\nu_x(ua)}$ for any word u. Hence,

$$\theta(x_1 \dots x_s) \le \prod_{u\in A^m} \prod_{a\in A} \theta(a/u)^{\nu_x(ua)}$$

$$\le \prod_{u\in A^m} \prod_{a\in A} (\nu_x(ua)/\bar{\nu}_x(u))^{\nu_x(ua)}.$$

If we apply those inequalities to $\theta(x^1 \diamond \dots \diamond x^r)$, we immediately obtain the following inequalities

$$\theta(x^1 \diamond \dots \diamond x^r) \le \prod_{u\in A^m} \prod_{a\in A} \theta(a/u)^{\nu_{x^1\diamond\dots\diamond x^r}(ua)} \le$$

$$\prod_{u\in A^m} \prod_{a\in A} (\nu_{x^1\diamond\dots\diamond x^r}(ua)/\bar{\nu}_{x^1\diamond\dots\diamond x^r}(u))^{\nu_{x^1\diamond\dots\diamond x^r}(ua)}.$$

Now the statement of the Lemma 1.1 follows from the definition (1.34).

Proof (Theorem 1.6) Let C_α be a critical set of the test $T_\varphi^{id}(A,\alpha)$, i.e., by definition, $C_\alpha = \{u : u \in A^t \;\&\; -\log \pi(u) - |\varphi(u)| > -\log \alpha\}$. Let μ_φ be a measure for which the claim is true. We define an auxiliary set $\hat{C}_\alpha = \{u : -\log \pi(u) - (-\log \mu_\varphi(u)) > -\log \alpha\}$. We have $1 \geq \sum_{u\in\hat{C}_\alpha} \mu_\varphi(u) \geq \sum_{u\in\hat{C}_\alpha} \pi(u)/\alpha = (1/\alpha)\pi(\hat{C}_\alpha)$. (Here the second inequality follows from the definition of $\hat{C}_\alpha$, whereas all others are obvious.) So, we obtain that $\pi(\hat{C}_\alpha) \leq \alpha$. From definitions of $C_\alpha, \hat{C}_\alpha$ and (1.26) we immediately obtain that $\hat{C}_\alpha \supset C_\alpha$. Thus, $\pi(C_\alpha) \leq \alpha$. By definition, $\pi(C_\alpha)$ is the value of the Type I error. The first statement of the theorem is proven.

Let us prove the second statement of the theorem. Suppose that the hypothesis $H_1^{id}(A)$ is true. That is, the sequence $x_1 \dots x_t$ is generated by some stationary and ergodic source τ and $\tau \neq \pi$. Our strategy is to show that

$$\lim_{t\to\infty} -\log \pi(x_1 \dots x_t) - |\varphi(x_1 \dots x_t)| = \infty \tag{1.55}$$

with probability 1 (according to the measure τ). First we represent (1.55) as

$$-\log \pi(x_1 \dots x_t) - |\varphi(x_1 \dots x_t)|$$

$$= t(\frac{1}{t} \log \frac{\tau(x_1 \dots x_t)}{\pi(x_1 \dots x_t)} + \frac{1}{t}(-\log \tau(x_1 \dots x_t) - |\varphi(x_1 \dots x_t)|)).$$

From this equality and the property of a universal code (1.29) we obtain

$$-\log \pi(x_1 \dots x_t) - |\varphi(x_1 \dots x_t)| = t\,(\frac{1}{t} \log \frac{\tau(x_1 \dots x_t)}{\pi(x_1 \dots x_t)} + o(1)). \tag{1.56}$$

From (1.29) and (1.21) we can see that

$$\lim_{t\to\infty} -\log \tau(x_1 \dots x_t)/t \leq h_k(\tau) \tag{1.57}$$

for any $k \geq 0$ (with probability 1). It is supposed that the process π has a finite memory, i.e., belongs to $M_s(A)$ for some s. Having taken into account the definition of $M_s(A)$ (1.18), we obtain the following representation:

$$-\log \pi(x_1 \dots x_t)/t = -t^{-1} \sum_{i=1}^{t} \log \pi(x_i/x_1 \dots x_{i-1})$$

$$= -t^{-1}(\sum_{i=1}^{k} \log \pi(x_i/x_1 \dots x_{i-1}) + \sum_{i=k+1}^{t} \log \pi(x_i/x_{i-k} \dots x_{i-1}))$$

for any $k \geq s$. According to the ergodic theorem there exists a limit

$$\lim_{t\to\infty} t^{-1} \sum_{i=k+1}^{t} \log \pi(x_i/x_{i-k} \dots x_{i-1}),$$

which is equal to $h_k(\tau)$ [4, 14]. So, from the two last equalities we can see that

$$\lim_{t\to\infty}(-\log\pi(x_1\dots x_t))/t = -\sum_{v\in A^k}\tau(v)\sum_{a\in A}\tau(a/v)\log\pi(a/v).$$

Taking into account this equality, (1.57) and (1.56), we can see that

$$-\log\pi(x_1\dots x_t) - |\varphi(x_1\dots x_t)| \geq$$
$$t\,(\sum_{v\in A^k}\tau(v)\sum_{a\in A}\tau(a/v)\log(\tau(a/v)/\pi(a/v))) + o(t)$$

for any $k \geq s$. From this inequality we can obtain that

$$-\log\pi(x_1\dots x_t) - |\varphi(x_1\dots x_t)| \geq c\,t + o(t),$$

where c is a positive constant, $t \to \infty$. Hence, (1.55) is true and the theorem is proven.

Proof (Theorem 1.7) Let us denote the critical set of the test $T_\varphi^{SI}(A,\alpha)$ as C_α, i.e., by definition, $C_\alpha = \{x_1\dots x_t :\ (t-m)\,h_m^*(x_1\dots x_t) - |\varphi(x_1\dots x_t)|) > \log(1/\alpha)\}$. From Claim 1.2 we can see that there exists such a measure μ_φ that $-\log\mu_\varphi(x_1\dots x_t) \leq |\varphi(x_1\dots x_t)|$. We also define

$$\hat{C}_\alpha = \{x_1\dots x_t :\ (t-m)\,h_m^*(x_1\dots x_t) - (-\log\mu_\varphi(x_1\dots x_t))\,) > \log(1/\alpha)\}. \tag{1.58}$$

Obviously, $\hat{C}_\alpha \supset C_\alpha$. Let θ be any source from $M_m(A)$. The following chain of equalities and inequalities is true:

$$1 \geq \mu_\varphi(\hat{C}_\alpha) = \sum_{x_1\dots x_t\in\hat{C}_\alpha}\mu_\varphi(x_1\dots x_t)$$
$$\geq \alpha^{-1}\sum_{x_1\dots x_t\in\hat{C}_\alpha}2^{(t-m)h_m^*(x_1\dots x_t)} \geq \alpha^{-1}\sum_{x_1\dots x_t\in\hat{C}_\alpha}\theta(x_1\dots x_t) = \theta(\hat{C}_\alpha).$$

(Here both equalities and the first inequality are obvious, the second and the third inequalities follow from (1.58) and the lemma 1.1, correspondingly.) So, we obtain that $\theta(\hat{C}_\alpha) \leq \alpha$ for any source $\theta \in M_m(A)$. Taking into account that $\hat{C}_\alpha \supset C_\alpha$, where C_α is the critical set of the test, we can see that the probability of the Type I error is not greater than α. The first statement of the theorem is proven.

The proof of the second statement will be based on some results of Information Theory. We obtain from (1.29) that for any stationary and ergodic p

$$\lim_{t\to\infty} t^{-1}|\varphi(x_1\dots x_t)| = h_\infty(p) \tag{1.59}$$

with probability 1. It can be seen from (1.23) that h^*_m is an estimate for the $m-$order Shannon entropy (1.20). Applying the ergodic theorem we obtain $\lim_{t\to\infty} h^*_m(x_1\dots x_t) = h_m(p)$ with probability 1 [4, 14]. It is known in Information Theory that $h_m(\varrho) - h_\infty(\varrho) > 0$ if ϱ belongs to $M_\infty(A) \setminus M_m(A)$ [4, 14]. It is supposed that H_1^{SI} is true, i.e., the considered process belongs to $M_\infty(A) \setminus M_m(A)$. So, from (1.59) and the last equality we obtain that $\lim_{t\to\infty}((t-m)\, h^*_m(x_1\dots x_t) - |\varphi(x_1\dots x_t)|) = \infty$. This proves the second statement of the theorem.

Proof (Theorem 1.8) First we prove that with probability 1 there exists the limit $\lim_{t\to\infty} \frac{1}{t}\log(p(x_1\dots x_t)/r_U(x_1\dots x_t))$ and this limit is finite and nonnegative. Let $A_n = \{x_1,\dots,x_n : p(x_1,\dots,x_n) \neq 0\}$. Define

$$z_n(x_1\dots x_n) = r_U(x_1\dots x_n)/p(x_1\dots x_n) \tag{1.60}$$

for $(x_1,\dots,x_n) \in A$ and $z_n = 0$ elsewhere.

Since

$$E_P(z_n|x_1,\dots,x_{n-1}) = E\left(\frac{r_U(x_1\dots x_n)}{p(x_1\dots x_n)}\middle| x_1,\dots,x_{n-1}\right)$$
$$= \frac{r_U(x_1\dots x_{n-1})}{p(x_1\dots x_{n-1})} E_P\left(\frac{r_U(x_n|x_1\dots x_{n-1})}{p(x_n|x_1\dots x_{n-1})}\right)$$
$$= z_{n-1}\int_A \frac{r_U(x_n|x_1\dots x_{n-1})dP(x_n|x_1\dots x_{n-1})}{dP(x_n|x_1\dots x_{n-1})/dM_n(x_n|x_1\dots x_{n-1})}f$$
$$= z_{n-1}\int_A r_U(x_n|x_1\dots x_{n-1})dM_n(x_n|x_1\dots x_{n-1}) \le z_{n-1}$$

the stochastic sequence (z_n, B^n) is, by definition, a non-negative supermartingale with respect to P, with $E(z_n) \le 1$; see [48]. Hence, Doob's submartingale convergence theorem implies that the limit z_n exists and is finite with P-probability 1 (see [48, Theorem 7.4.1]). Since all terms are nonnegative so is the limit. Using the definition (1.60) with P-probability 1 we have

$$\lim_{n\to\infty} p(x_1\dots x_n)/r_U(x_1\dots x_n) > 0,$$

$$\lim_{n\to\infty} \log(p(x_1\dots x_n)/r_U(x_1\dots x_n)) > -\infty$$

and

$$\lim_{n\to\infty} n^{-1}\log(p(x_1\dots x_n)/r_U(x_1\dots x_n)) \ge 0. \tag{1.61}$$

Now we note that for any integer s the following obvious equality is true: $r_U(x_1 \ldots x_t) = \omega_s \mu_U(x_1^{[s]} \ldots x_t^{[s]})/M_t(x_1^{[s]} \ldots x_t^{[s]})\ (1+\delta)$ for some $\delta > 0$. From this equality, (1.31) and (1.43) we immediately obtain that a.s.

$$\lim_{t\to\infty} \frac{1}{t} \log \frac{p(x_1 \ldots x_t)}{r_U(x_1 \ldots x_t)} \leq \lim_{t\to\infty} \frac{-\log \omega_t}{t}$$
$$+ \lim_{t\to\infty} \frac{1}{t} \log \frac{p(x_1 \ldots x_t)}{\mu_U(x_1^{[s]} \ldots x_t^{[s]})/M_t(x_1^{[s]} \ldots x_t^{[s]})}$$
$$\leq \lim_{t\to\infty} \frac{1}{t} \log \frac{p(x_1 \ldots x_t)}{2^{-|U(x_1^{[s]} \ldots x_t^{[s]})|}/M_t(x_1^{[s]} \ldots x_t^{[s]})}. \tag{1.62}$$

The right part can be presented as follows:

$$\lim_{t\to\infty} \frac{1}{t} \log \frac{p(x_1 \ldots x_t)}{2^{-|U(x_1^{[s]} \ldots x_t^{[s]})|}/M_t(x_1^{[s]} \ldots x_t^{[s]})}$$
$$= \lim_{t\to\infty} \frac{1}{t} \log \frac{p^s(x_1 \ldots x_t)\, M_t(x_1^{[s]} \ldots x_t^{[s]})}{2^{-|U(x_1^{[s]} \ldots x_t^{[s]})|}} \tag{1.63}$$
$$+ \lim_{t\to\infty} \frac{1}{t} \log \frac{p(x_1 \ldots x_t)}{p^s(x_1 \ldots x_t)}.$$

Having taken into account that U is a universal code, (1.40) and Theorem 1.2, we can see that the first term is equal to zero. From (1.39) and (1.42) we can see that a.s. the second term is equal to $\tilde{h}_s - \tilde{h}$. This equality is valid for any integer s and, according to (1.44), the second term equals zero, too, and we obtain that

$$\lim_{t\to\infty} \frac{1}{t} \log \frac{p(x_1 \ldots x_t)}{r_U(x_1 \ldots x_t)} \leq 0.$$

Having taken into account (1.61), we can see that the first statement is proven.

From (1.62) and (1.63) we can see that

$$E \log \frac{p(x_1 \ldots x_t)}{r_U(x_1 \ldots x_t)} \leq E \log \frac{p_t^s(x_1, \ldots, x_t)\, M_t(x_1^{[s]} \ldots x_t^{[s]})}{2^{-|U(x_1^{[s]} \ldots x_t^{[s]})|}} \tag{1.64}$$
$$+ E \log \frac{p(x_1 \ldots x_t)}{p^s(x_1, \ldots, x_t)}.$$

The first term is the average redundancy of the universal code for a finite-alphabet source; hence, according to Theorem 1.2, it tends to 0. The second term tends to $\tilde{h}_s - \tilde{h}$ for any s and from (1.44) we can see that it is equals to zero. The second statement is proven.

Proof (Theorem 1.9) Obviously,

$$E(\frac{1}{t}\sum_{m=0}^{t-1}(P(x_{m+1} \in B_{m+1}|x_1 \dots x_m) - R_U(x_{m+1} \in B_{m+1}|x_1 \dots x_m))^2) \le \quad (1.65)$$

$$\frac{1}{t}\sum_{m=0}^{t-1} E(|P(x_{m+1} \in B_{m+1}|x_1 \dots x_m) - R_U(x_{m+1} \in B_{m+1}|x_1 \dots x_m)| +$$

$$|P(x_{m+1} \in \bar{B}_{m+1}|x_1 \dots x_m) - R_U(x_{m+1} \in \bar{B}_{m+1}|x_1 \dots x_m)|)^2.$$

From the Pinsker inequality (1.5) and convexity of the KL divergence (1.6) we obtain the following inequalities:

$$\frac{1}{t}\sum_{m=0}^{t-1} E(|P(x_{m+1} \in B_{m+1}|x_1 \dots x_m) - R_U(x_{m+1} \in B_{m+1}|x_1 \dots x_m)| + \quad (1.66)$$

$$|P(x_{m+1} \in \bar{B}_{m+1}|x_1 \dots x_m) - R_U(x_{m+1} \in \bar{B}_{m+1}|x_1 \dots x_m)|)^2 \le$$

$$\frac{const}{t}\sum_{m=0}^{t-1} E((\log \frac{P(x_{m+1} \in B_{m+1}|x_1 \dots x_m)}{R_U(x_{m+1} \in B_{m+1}|x_1 \dots x_m)} + \log \frac{P(x_{m+1} \in \bar{B}_{m+1}|x_1 \dots x_m)}{R_U(x_{m+1} \in \bar{B}_{m+1}|x_1 \dots x_m)}) \le$$

$$\frac{const}{t}\sum_{m=0}^{t-1}(\int p(x_1 \dots x_m)(\int p(x_{m+1}|x_1 \dots x_m)) \log \frac{p(x_{m+1}|x_1 \dots x_m)}{r_U(x_{m+1}|x_1 \dots x_m)} dM) dM_m).$$

Having taken into account that the last term is equal to $\frac{const}{t} E(\log \frac{p(x_1 \dots x_t)}{r_U(x_1 \dots x_t)})$, from (1.65), (1.66) and (1.46) we obtain (1.47). ii) can be derived from i) and the Jensen inequality for the function x^2.

Proof (Theorem 1.10) The last inequality of the following chain follows from Pinsker's, whereas all others are obvious.

$$(\int f(x)\, p(x|x_1 \dots x_m)\, dM_m - \int f(x)\, r_U(x|x_1 \dots x_m)\, dM_m)^2$$

$$= (\int f(x)\, (p(x|x_1 \dots x_m) - r_U(x|x_1 \dots x_m))\, dM_m)^2$$

$$\le \bar{M}^2(\int (p(x|x_1 \dots x_m) - r_U(x|x_1 \dots x_m))\, dM_m)^2$$

$$\le \bar{M}^2(\int |p(x|x_1 \dots x_m) - r_U(x|x_1 \dots x_m)| dM_m)^2$$

$$\le const \int p(x|x_1 \dots x_m) \log \frac{p(x|x_1 \dots x_m)}{r_U(x|x_1 \dots x_m)} dM_m.$$

From these inequalities we obtain:

$$E(\sum_{m=0}^{t-1}(\int f(x)\,p(x|x_1\dots x_m)\,dM_m -$$

$$\int f(x)\,r_U(x|x_1\dots x_m)\,dM_m)^2) \le \tag{1.67}$$

$$\sum_{m=0}^{t-1} const\,E(\int p(x|x_1\dots x_m)\,\log\frac{p(x|x_1\dots x_m)}{r_U(x|x_1\dots x_m)}dM_{1/m}).$$

The last term can be presented as follows:

$$\sum_{m=0}^{t-1} E(\int p(x|x_1\dots x_m)\,\log\frac{p(x|x_1\dots x_m)}{r_U(x|x_1\dots x_m)}dM_{1/m}) =$$

$$\sum_{m=0}^{t-1}\int p(x_1\dots x_m)$$

$$\int p(x|x_1\dots x_m)\log\frac{p(x|x_1\dots x_m)}{r_U(x|x_1\dots x_m)}dM_{1/m}dM_m$$

$$= \int p(x_1\dots x_t)\log(p(x_1\dots x_t)/r_U(x_1\dots x_t))dM_t.$$

From this equality and (1.67) we obtain (1.48). ii) can be derived from (1.67) and the Iensen inequality for the function x^2.

Proof (Theorem 1.11) The following chain proves the first statement of the theorem:

$$P\{H_0^{\aleph}(A)\ is\ rejected\ /H_0\ is\ true\} = P\{\bigcup_{i=1}^{\infty}\{H_0^{\aleph}(\Lambda_i)\ is\ rejected\ /H_0\ is\ true\}\,\}$$

$$\le \sum_{i=1}^{\infty} P\{H_0^{\aleph}(\Lambda_i)\ /H_0\ is\ true\} \le \sum_{i=1}^{\infty}(\alpha\omega_i) = \alpha.$$

(Here both inequalities follow from the description of the test, whereas the last equality follows from (1.24).)

The second statement also follows from the description of the test. Indeed, let a sample be created by a source ϱ, for which $H_1(A)^{\aleph}$ is true. It is supposed that the sequence of partitions $\hat{\Lambda}$ discriminates between $H_0^{\aleph}(A), H_1^{\aleph}(A)$. By definition, it means that there exists j for which $H_1^{\aleph}(\Lambda_j)$ is true for the process ϱ_{Λ_j}. It immediately follows from Theorems 1.1–1.4 that the Type II error of the test $T_{\varphi}^{\aleph}(\Lambda_j, \alpha\omega_j)$ goes to 0 when the sample size tends to infinity.

References

1. Algoet, P.: Universal schemes for learning the best nonlinear predictor given the infinite past and side information. IEEE Trans. Inf. Theory **45**, 1165–1185 (1999)
2. Babu, G.J., Boyarsky, A., Chaubey, Y.P., Gora, P.: New statistical method for filtering and entropy estimation of a chaotic map from noisy data. Int. J. Bifurc. Chaos **14** (11), 3989–3994 (2004)
3. Barron, A.R.: The strong ergodic theorem for densities: generalized Shannon-McMillan-Breiman theorem. Ann. Probab. **13** (4), 1292–1303 (1985)
4. Billingsley, P.: Ergodic Theory and Information. Wiley, New York (1965)
5. Board of Governors of the Federal Reserve System (U.S.): http://www.federalreserve.gov/releases/g17/current/ (2012)
6. Cilibrasi, R., Vitanyi, P.M.B.: Clustering by compression. IEEE Trans. Inf. Theory **51**(4), 1523–1545 (2005)
7. Cilibrasi, R., de Wolf, R., Vitanyi, P.M.B.: Algorithmic clustering of music. Comput. Music J. **28**(4), 49–67 (2004)
8. Csiszár, I., Shields, P.C.: Information Theory and Statistics: A Tutorial. In: Foundations and Trends in Communications and Information Theory. Now Publishers Inc., Hanover (2004)
9. Csiszár, I., Shields, P.C.: The consistency of the BIC Markov order estimation. Ann. Stat. **6**, 1601–1619 (2000)
10. Effros, M., Visweswariah, K., Kulkarni, S.R., Verdu, S.: Universal lossless source coding with the Burrows–Wheeler transform. IEEE Trans. Inf. Theory **45**, 1315–1321 (1999)
11. Feller, W.: An Introduction to Probabability Theory and Its Applications, vol. 1. Wiley, New York (1970)
12. Finesso, L., Liu, C., Narayan, P.: The optimal error exponent for Markov order estimation. IEEE Trans. Inf. Theory **42**, 1488–1497 (1996)
13. Fitingof, B.M.: Optimal encoding for unknown and changing statistics of messages. Probl. Inf. Transm. **2**(2), 3–11 (1966)
14. Gallager, R.G.: Information Theory and Reliable Communication. Wiley, New York (1968)
15. Györfi, L., Páli, I., van der Meulen, E.C.: There is no universal code for infinite alphabet. IEEE Trans. Inf. Theory **40**, 267–271 (1994)
16. International Institute of Forecasters: http://forecasters.org/resources/time-series-data/ (2011)
17. Jevtic, N., Orlitsky, A., Santhanam, N.P.: A lower bound on compression of unknown alphabets. Theor. Comput. Sci. **332**, 293–311 (2004)
18. Kelly, J.L.: A new interpretation of information rate. Bell Syst. Tech. J. **35**, 917–926 (1956)
19. Kieffer, J.: A unified approach to weak universal source coding. IEEE Trans. Inf. Theory **24**, 674–682 (1978)
20. Kieffer, J.: Prediction and Information Theory. Preprint, available at ftp://oz.ee.umn.edu/users/kieffer/papers/prediction.pdf/ (1998)
21. Kieffer, J.C., Yang, E.-H.: Grammar-based codes: a new class of universal lossless source codes. IEEE Trans. Inf. Theory **46**(3), 737–754 (2000)
22. Kolmogorov, A.N.: Three approaches to the quantitative definition of information. Probl. Inf. Transm. **1**, 3–11 (1965)
23. Knuth, D.E.: The Art of Computer Programming, vol. 2. Addison–Wesley, Reading (1981)
24. Krichevsky, R.: A relation between the plausibility of information about a source and encoding redundancy. Probl. Inf. Transm. **4**(3), 48–57 (1968)
25. Krichevsky, R.: Universal Compression and Retrival. Kluwer Academic, Dordecht (1993)
26. Kullback, S.: Information Theory and Statistics. Wiley, New York (1959)
27. Modha, D.S., Masry, E.: Memory-universal prediction of stationary random processes. IEEE Trans. Inf. Theory **44**(1), 117–133 (1998)
28. Nobel, A.B.: On optimal sequential prediction. IEEE Trans. Inf. Theory **49**(1), 83–98 (2003)
29. Orlitsky, A., Santhanam, N.P., Zhang, J.: Always good turing: asymptotically optimal probability estimation. Science **302**, 427–431 (2003)

30. Rissanen, J.: Generalized Kraft inequality and arithmetic coding. IBM J. Res. Dev. **20**(5), 198–203 (1976)
31. Rissanen, J.: Universal coding, information, prediction, and estimation. IEEE Trans. Inf. Theory **30**(4), 629–636 (1984)
32. Ryabko, B.Ya.: Twice-universal coding. Probl. Inf. Transm. **20**(3), 173–177 (1984)
33. Ryabko, B.Ya.: Prediction of random sequences and universal coding. Probl. Inf. Transm. **24**(2), 87–96 (1988)
34. Ryabko, B.Ya.: A fast adaptive coding algorithm. Probl. Inf. Transm. **26**(4), 305–317 (1990)
35. Ryabko, B.Ya.: The complexity and effectiveness of prediction algorithms. J. Complex. **10**(3), 281–295 (1994)
36. B. Ryabko, Astola, J.: Universal codes as a basis for time series testing. Stat. Methodol. **3**, 375–397 (2006)
37. B. Ryabko, Astola, J., Gammerman, A.: Application of Kolmogorov complexity and universal codes to identity testing and nonparametric testing of serial independence for time series. Theor. Comput. Sci. **359**, 440–448 (2006)
38. Ryabko, B., Astola, J., Gammerman, A.: Adaptive coding and prediction of sources with large and infinite alphabets. IEEE Trans. Inf. Theory **54**(8), 440–448 (2008)
39. B. Ryabko, Astola, J., Egiazarian, K.: Fast codes for large alphabets. Commun. Inf. Syst. **3** (2), 139–152 (2003)
40. B. Ryabko, Monarev, V.: Experimental investigation of forecasting methods based on data compression algorithms. Probl. Inf. Transm. **41**(1), 65–69 (2005)
41. Ryabko, B., Reznikova, Zh.: Using Shannon entropy and Kolmogorov complexity to study the communicative system and cognitive capacities in ants. Complexity **2**(2), 37–42 (1996)
42. Ryabko, B., Reznikova, Zh.: The use of ideas of information theory for studying "Language" and intelligence in ants. Entropy **11**(4), 836–853 (2009)
43. Ryabko, B., Topsoe, F.: On asymptotically optimal methods of prediction and adaptive coding for Markov sources. J. Complex. **18**(1), 224–241 (2002)
44. Ryabko, D., Hutter, M.: Sequence prediction for non-stationary processes. In: Proceedings: Combinatorial and Algorithmic Foundations of Pattern and Association Discovery, Dagstuhl Seminar, 2006. http://www.dagstuhl.de/06201/, see also http://arxiv.org/pdf/cs.LG/0606077 (2006)
45. Savari, S.A.: A probabilistic approach to some asymptotics in noiseless communication. IEEE Trans. Inf. Theory **46**(4), 1246–1262 (2000)
46. Shannon, C.E.: A mathematical theory of communication. Bell Syst. Tech. J. **27**, 379–423, 623–656 (1948)
47. Shannon, C.E.: Communication theory of secrecy systems. Bell Syst. Tech. J. **28**, 656–715 (1948)
48. Shiryaev, A.N.: Probability, 2nd edn. Springer, New York (1995)

Chapter 2
Applications to Cryptography

2.1 Introduction

In this part we consider the problem of detecting deviations of binary sequence from randomness in details, because this problem is very important for cryptography for the following reasons.

Random number (RNG) and pseudorandom number generators (PRNG) are widely used in cryptography and more generally in data security systems. That is why statistical tests for detecting deviations are of great interest for cryptography [26, 27, 33]. (Thus, the National Institute of Standards and Technology (NIST, USA) has elaborated "A statistical test suite for random and pseudorandom number generators for cryptographic applications" [33].)

Stream ciphers and block ciphers are very important tools of cryptography, and both of them are connected with PRNG. More precisely, a stream cipher generates a so-called keystream which, in fact, is a bit-sequence which must be similar to the sequence obtained by tossing a perfect coin. The block ciphers are used for solving many problems in cryptography and there are two modes of their usage where they used as PRNG. (Namely, OFB and CTR modes, see [21, 28, 37].) So, if any test detects a deviation from randomness of sequences generated by a certain cipher, it will be considered as a sufficient drawback of the cipher, see Knudsen and Meier (Correlation in RC6, private communication. http://www.ii.uib.no/larsr/papers/rc6.ps) [21, 33, 41].

There exists a so-called gradient statistical attack on block ciphers which is based on detecting deviations of binary sequences. This attack will be described at the end of this chapter.

In this chapter we consider the tests for detecting deviations from randomness and show how common data compression methods can be used as the test. Despite the fact that this problem is a particular case of the above considered goodness-of-fit test, we describe it in detail because, first, it is very important for applications and, second, there exist simple and efficient approaches which cannot be used in

B. Ryabko et al., *Compression-Based Methods of Statistical Analysis and Prediction of Time Series*, DOI 10.1007/978-3-319-32253-7_2

a general case. Besides, we describe two tests which are based on some ideas of universal codes and apply them to a real PRNG and for a real stream cipher. It turns out that in both cases the described tests detected deviations from randomness. The last part of this chapter is devoted to application of the described tests to the attack on block ciphers.

The randomness testing of random number and pseudorandom number generators is used for many purposes including cryptographic, modeling and simulation applications; see, for example, [19, 23, 28]. For such applications a required bit sequence should be truly random, i.e., by definition, such a sequence could be interpreted as the result of the flips of a fair coin with sides that are labeled 0 and 1 (for short, it is called a random sequence; see [33]). More formally, we will consider the main hypothesis H_0 that a bit sequence is generated by the Bernoulli source with equal probabilities of 0's and 1's. Associated with this null hypothesis is the alternative hypothesis H_1 that the sequence is generated by a stationary and ergodic source which generates letters from $\{0, 1\}$ and differs from the source under H_0.

In this part we will consider some tests which are based on results and ideas of Information Theory and, in particular, source coding theory. First, we show that a universal code can be used for randomness testing. If we take into account that the Shannon per-bit entropy is maximal (1 bit) if H_0 is true and is less than 1 if H_1 is true, we see that it is natural to use this property and universal codes for randomness testing because, in principle, such a test can distinguish each deviation from randomness, which can be described in a framework of the stationary and ergodic source model. Loosely speaking, the test rejects H_0 if a binary sequence can be compressed by a considered universal code (or a data compression method).

It should be noted that the idea to use the compressibility as a measure of randomness has a long history in mathematics. The point is that, on the one hand, the problem of randomness testing is quite important in practice, but, on the other hand, this problem is closely connected with such deep theoretical issues as the definition of randomness, the logical basis of probability theory, randomness and complexity, etc; see [4, 19, 20, 24]. Thus, Kolmogorov [20] suggested defining the randomness of a sequence, informally, as the length of the shortest program, which can create the sequence (if one of the universal Turing machines is used as a computer). So, loosely speaking, the randomness (or Kolmogorov complexity) of the finite sequence is equal to its shortest description. It is known that the Kolmogorov complexity is not computable and, therefore, cannot be used for randomness testing. On the other hand, each lossless data compression code can be considered as a method for upper bounding the Kolmogorov complexity. Indeed, if x is a binary word, ϕ is a data compression code and $\phi(x)$ is the codeword of x, then the length of the codeword $|\phi(x)|$ is the upper bound for the Kolmogorov complexity of the word x. So, again we see that the codeword length of the lossless data compression method can be used for randomness testing.

In this chapter we proceed with a consideration of the tests for randomness, which are based on results and ideas from source coding theory. Firstly, we show how to build a test based on any data compression method and give some examples of application of such a test to PRNG testing. It should be noted that data compression methods were considered as a basis for randomness testing in the literature. For example, Maurer's Universal Statistical Test, the Lempel-Ziv Compression Test and the Approximate Entropy Test are connected with universal codes, see, for example, [33]. In contrast to known methods, our approach gives a possibility to make a test for randomness, basing it on any lossless data compression method even if a distribution law of the codeword lengths is not known.

Secondly, in this part, we describe two new tests, conceptually connected with universal codes. When both tests are applied, a tested sequence $x_1x_2\dots x_n$ is divided into subwords $x_1x_2\dots x_s$, $x_{s+1}x_{s+2}\dots x_{2s}$, ..., $s \geq 1$, and the hypothesis H_0^* that the subwords obey the uniform distribution (i.e., each subword is generated with probability 2^{-s}) is tested against $H_1^* = \neg H_0^*$. The key idea of the new tests is as follows. All subwords from the set $\{0,1\}^s$ are ordered and this order changes after processing each subword $x_{js+1}x_{js+2}\dots x_{(j+1)s}$, $j = 0, 1, \dots,$ in such a way that, loosely speaking, the more frequent subwords have small ordinals. When the new tests are applied, the frequency of different ordinals are estimated (instead of frequencies of the subwords as for, say, the chi-square test).

The natural question is how to choose the block length s in such schemes. We show that, informally speaking, the block length s should be taken quite large due to the existence of so-called *two-faced processes*. More precisely, it is shown that for each integer s^* there exists such a process ξ that for each binary word u the process ξ creates u with probability $2^{-|u|}$ if the length of u ($|u|$) is less than or equal to s^*, but, on the other hand, the probability distribution $\xi(v)$ is very far from uniform if the length of the words v is greater than s^*. (So, if we use a test with the block length $s \leq s^*$, the sequences generated by ξ will look random, in spite of ξ being far from random.)

2.2 Data Compression Methods as a Basis for Randomness Testing

2.2.1 Randomness Testing Based on Data Compression

Let A be a finite alphabet and A^n be the set of all words of length n over A, where n is an integer. By definition, $A^* = \bigcup_{n=1}^{\infty} A^n$ and A^{∞} is the set of all infinite words $x_1x_2\dots,$ over the alphabet A. A data compression method (or code) φ is defined as a set of mappings φ_n such that $\varphi_n : A^n \to \{0,1\}^*$, $n = 1, 2, \dots$ and for each pair of different words $x, y \in A^n$ $\varphi_n(x) \neq \varphi_n(y)$. Informally, it means that the code φ can be applied for compression of each message of any length $n, n > 0$, over alphabet A and the message can be decoded if its code is known.

Now we can describe a statistical test which can be constructed based on any code φ. Let n be an integer and $\hat{H}_0$ be a hypothesis that the words from the set A^n obey the uniform distribution, i.e., $p(u) = |A|^{-n}$ for each $u \in \{0,1\}^n$. (Here and below $|x|$ is the length of x if x is a word, and the number of elements in x if x is a set.) Let a required level of significance (or a Type I error) be α, $\alpha \in (0, 1)$. The following main idea of a suggested test is quite natural: The well-compressed words should be considered as non-random and $\hat{H}_0$ should be rejected. More exactly, we define a critical value of the suggested test by

$$t_\alpha = n \log |A| - \log(1/\alpha) - 1\,. \tag{2.1}$$

(Here and below $\log x = \log_2 x$.)

Let u be a word from A^n. By definition, the hypothesis $\hat{H}_0$ is accepted if $|\varphi_n(u)| > t_\alpha$ and rejected if $|\varphi_n(u)| \le t_\alpha$. We denote this test by $\Gamma^{(n)}_{\alpha,\varphi}$.

Theorem 2.1 *For each integer n and a code φ, the Type I error of the described test $\Gamma^{(n)}_{\alpha,\varphi}$ is not larger than α.*

Proof is given in Sect. 2.6.4.

Comment 2.1 The described test can be modified in such a way that the Type I error will be equal to α. For this purpose we define the set A_γ by

$$A_\gamma = \{x : x \in A^n \;\; \& \;\; |\varphi_n(x)| = \gamma\}$$

and an integer g for which the two following inequalities are valid:

$$\sum_{j=0}^{g} |A_j| \le \alpha |A|^n < \sum_{j=0}^{g+1} |A_j|\,. \tag{2.2}$$

Now the modified test can be described as follows:

If for $x \in A^n$ $|\varphi_n(x)| \le g$ then $\hat{H}_0$ is rejected, if $|\varphi_n(x)| > (g+1)$ then $\hat{H}_0$ is accepted and if $|\varphi_n(x)| = (g+1)$ the hypothesis $\hat{H}_0$ is accepted with the probability

$$(\sum_{j=1}^{g+1} |A_j| - \alpha |A|^n)/|A_{g+1}|$$

and rejected with the probability

$$1 - (\sum_{j=1}^{g+1} |A_j| - \alpha |A|^n)/|A_{g+1}|\,.$$

(Here we used a randomized criterion, see for definition, for example, [18], part 22.11.) We denote this test by $\Upsilon^{(n)}_{\alpha,\varphi}$.

Claim 2.1 For each integer n and a code φ, the Type I error of the described test $\Upsilon^{(n)}_{\alpha,\varphi}$ is equal to α.

Proof is given in Sect. 2.6.4.

We can see that this criterion has the level of significance (or Type I error) exactly α, whereas the first criterion, which is based on critical value (2.1), has the level of significance that could be less than α. In spite of this drawback, the first criterion may be more useful due to its simplicity. Moreover, such an approach gives a possibility to use a data compression method ψ for testing even in the case where the distribution of the length $|\psi_n(x)|, x \in A^n$ is not known.

Comment 2.2 We have considered codes for which different words of the same length have different codewords. (In Information Theory sometimes such codes are called non-singular.) Quite often a stronger restriction is required in Information Theory. Namely, it is required that each sequence $\varphi_n(x_1)\varphi_n(x_2)\ldots\varphi(x_r), r \geq 1$, of encoded words from the set $A^n, n \geq 1$, be uniquely decodable into $x_1x_2\ldots x_r$. Such codes are called uniquely decodable. For example, let $A = \{a, b\}$; the code $\psi_1(a) = 0, \psi_1(b) = 00$, obviously, is non-singular, but is not uniquely decodable. (Indeed, the word 000 can be decoded in both *ab* and *ba*.) It is well known in Information Theory that a code φ can be uniquely decoded if the following Kraft inequality is valid:

$$\Sigma_{u \in A^n} 2^{-|\varphi_n(u)|} \leq 1 \; ; \tag{2.3}$$

see, for example, [5, 13]. If it is known that the code is uniquely decodable, the suggested critical value (2.1) can be changed. Let us define

$$\hat{t}_\alpha = n \log |A| - \log(1/\alpha) \,. \tag{2.4}$$

Let, as before, u be a word from A^n. By definition, the hypothesis $\hat{H}_0$ is accepted if $|\varphi_n(u)| > \hat{t}_\alpha$ and rejected if $|\varphi_n(u)| \leq \hat{t}_\alpha$. We denote this test by $\hat{\Gamma}^{(n)}_{\alpha,\varphi}$.

Claim 2.2 For each integer n and a uniquely decodable code φ, the Type I error of the described test $\hat{\Gamma}^{(n)}_{\alpha,\varphi}$ is not larger than α.

Proof is given in Sect. 2.6.4.

So, we can see from (2.1) and (2.4) that the critical value is larger, if the code is uniquely decodable. On the other hand, the difference is quite small and (2.1) can be used without a large lose of the test power even in the case of uniquely decodable codes.

It should not be a surprise that the level of significance (or a Type I error) does not depend on the alternative hypothesis H_1, but, of course, the power of a test (and the Type II error) will be determined by H_1.

The examples of testing by real data compression methods are given in Sect. 2.4.

2.2.2 *Randomness Testing Based on Universal Codes*

We will consider the main hypothesis H_0 that the letters of a given sequence $x_1x_2\ldots x_t$, $x_i \in A$, are independent and identically distributed (i.i.d.) with equal probabilities of all $a \in A$ and the alternative hypothesis H_1 that the sequence is generated by a stationary and ergodic source, which generates letters from A and differs from the source under H_0. (If $A = \{0, 1\}$, i.i.d. coincides with Bernoulli source.) The definition of the stationary and ergodic source can be found in [3, 5, 13].

We will consider statistical tests, which are based on universal coding and universal prediction. First we define a universal code.

By definition, φ is a universal code if for each stationary and ergodic source (or a process) π the following equality is valid with probability 1 (according to the measure π):

$$\lim_{n\to\infty} (|\varphi_n(x_1\ldots x_n)|)/n = h(\pi)\,, \tag{2.5}$$

where $h(\pi)$ is the Shannon entropy. It is well known in Information Theory that $h(\pi) = \log|A|$ if H_0 is true, and $h(\pi) < \log|A|$ if H_1 is true, see, for example, [3, 5, 13]. From this property and (2.5) we can easily arrive at the following theorem.

Theorem 2.2 *Let φ be a universal code, $\alpha \in (0, 1)$ be a level of significance and a sequence $x_1x_2\ldots x_n$, $n \geq 1$, be generated by a stationary ergodic source π. If the test $\Gamma^{(n)}_{\alpha,\varphi}$ described above is applied for testing H_0 (against H_1), then, with probability 1, the Type I error is not larger than α, and the Type II error goes to 0 when $n \to \infty$.*

So, we can see that each good universal code can be used as a basis for randomness testing. But the converse proposition is not true. Let, for example, there be a code whose codeword length is asymptotically equal to $(0.5 + h(\pi)/2)$ for each source π (with probability 1, where, as before, $h(\pi)$ is the Shannon entropy). This code is not good, because its codeword length does not trend to the entropy, but, obviously, such a code could be used as a basis for a test of randomness. So, informally speaking, the set of tests is larger than the set of universal codes.

2.3 Two New Tests for Randomness and Two-Faced Processes

Firstly, we suggest two tests which are based on ideas of universal coding, but they are described in such a way that can be understood without any knowledge of Information Theory.

2.3.1 The "Book Stack" Test

Let, as before, there be given an alphabet $A = \{a_1, \dots, a_S\}$, a source which generates letters from A, and two following hypotheses: the source is i.i.d. and $p(a_1) = \dots = p(a_S) = 1/S$ (H_0) and $H_1 = \neg H_0$. We should test the hypotheses basing them on a sample $x_1 x_2 \dots x_n$, $n \geq 1$, generated by the source. When the "book stack" test is applied, all letters from A are ordered from 1 to S and this order is changed after observing each letter x_t according to the formula

$$\nu^{t+1}(a) = \begin{cases} 1, & if x_t = a\,; \\ \nu^t(a) + 1, & if \nu^t(a) < \nu^t(x_t); \\ \nu^t(a), & if \nu^t(a) > \nu^t(x_t)\,, \end{cases} \tag{2.6}$$

where ν^t is the order after observing $x_1 x_2 \dots x_t$, $t = 1, \dots, n$, ν^1 is defined arbitrarily. (For example, we can define $\nu^1 = \{a_1, \dots, a_S\}$.) Let us explain (2.6) informally. Suppose that the letters of A make a stack, like a stack of books, and $\nu^1(a)$ is a position of a in the stack. Let the first letter x_1 of the word $x_1 x_2 \dots x_n$ be a. If it takes the i_1-th position in the stack ($\nu^1(a) = i_1$), then take a out of the stack and put it on the top. (It means that the order is changed according to (2.6).) Repeat the procedure with the second letter x_2 and the stack obtained, etc.

It can help us understand the main idea of the suggested method if we take into account that, if H_1 is true, then frequent letters from A (as frequently used books) will have relatively small numbers (will spend more time next to the top of the stack). On the other hand, if H_0 is true, the probability of finding each letter x_i at each position j is equal to $1/S$.

Let us proceed with the description of the test. The set of all indexes $\{1, \dots, S\}$ is divided into r, $r \geq 2$, subsets $A_1 = \{1, 2, \dots, k_1\}, A_2 = \{k_1 + 1, \dots, k_2\}, \dots, A_r = \{k_{r-1} + 1, \dots, k_r\}$. Then, using $x_1 x_2 \dots x_n$, we calculate how many $\nu^t(x_t)$, $t = 1, \dots, n$, belong to a subset A_k, $k = 1, \dots, r$. We define this number as n_k (or, more formally, $n_k = |\{t : \nu^t(x_t) \in A_k, t = 1, \dots, n\}|, k = 1, \dots, r$). Obviously, if H_0 is true, the probability of the event $\nu^t(x_t) \in A_k$ is equal to $|A_j|/S$. Then, using a "common" chi-square test we test the hypothesis $\hat{H}_0 = P\{\nu^t(x_t) \in A_k\} = |A_k|/S$ basing on the empirical frequencies $n_1, \dots, n_r$ against $\hat{H}_1 = \neg\hat{H}_0$. Let us recall that the value

$$x^2 = \sum_{i=1}^{r} \frac{(n_i - n(|A_i|/S))^2}{n(|A_i|/S)} \tag{2.7}$$

is calculated when the chi-square test is applied, see, for example, [18]. It is known that x^2 asymptotically follows the χ^2-square distribution with $(k - 1)$ degrees of freedom (χ^2_{k-1}) if $\hat{H}_0$ is true. If the level of significance (or a Type I error) of the χ^2 test is $\alpha, \alpha \in (0, 1)$, the hypothesis $\hat{H}_0$ is accepted when x^2 from (2.7) is less than the $(1 - \alpha)$ *value* of the χ^2_{k-1} distribution [18].

We do not describe the exact rule of how to construct the subsets $\{A_1, A_2, \ldots, A_r\}$, but we recommend implementing some experiments for finding the parameters which make the sample size minimal (or, at least, acceptable). The point is that there are many cryptographic and other applications where it is possible to implement some experiments for optimizing the parameter values and, then, to test a hypothesis based on independent data. For example, in case of testing a PRNG it is possible to seek suitable parameters using a part of generated sequence and then to test the PRNG using a new part of the sequence.

Let us consider a small example. Let $A = \{a_1, \ldots, a_6\}$, $r = 2$, $A_1 = \{a_1, a_2, a_3\}$, $A_2 = \{a_4, a_5, a_6\}$, $x_1 \ldots x_8 = a_3 a_6\, a_3\, a_3 a_6\, a_1\, a_6 a_1$. If $\nu_1 = 1, 2, 3, 4, 5, 6$, then $\nu_2 = 3, 1, 2, 4, 5, 6$, $\nu_3 = 6, 3, 1, 2, 4, 5$, etc., and $n_1 = 7, n_2 = 1$. We can see that the letters a_3 and a_6 are quite frequent and the book "stack" indicates this nonuniformity quite well. (Indeed, the average values of n_1 and n_2 equal 4, whereas the real values are 7 and 1, correspondingly.)

Examples of practical applications of this test will be given in Sect. 2.4, but here we make two notes. Firstly, we pay attention to the complexity of this algorithm. The "naive" method of transformation according to (2.6) could take the number of operations proportional to S, but there exist algorithms which can perform all operations in (2.6) using $O(\log S)$ operations. Such algorithms can be based on AVL-trees; see, for example, [1].

The last comment concerns the name of the method. The "book stack" structure is quite popular in Information Theory and Computer Science. In Information Theory this structure was first suggested as a basis of a universal code by Ryabko [34] and was rediscovered by Bentley, Sleator, Tarjan and Wei in 1986 [2], and Elias in 1987 [11] (see also a comment in [35] about a history of this code). Nowadays this code is frequently called the "Move-to-Front" (MTF) scheme, as was suggested in [2]. Besides, this data structure is used in so-called caching and many other algorithms in Computer Science under the name "Move-to-Front". It is also worth noting that the book stack was first considered by Soviet mathematician M.L. Cetlin as an example of a self-adaptive system in the 1960s, see [32].

2.3.2 The Order Test

This test is also based on changing the order $\nu^t(a)$ of alphabet letters but the rule of the order change differs from (2.6). To describe the rule we first define $\lambda^{t+1}(a)$ as a count of occurrences of a in the word $x_1 \ldots x_{t-1} x_t$. At each moment t the alphabet letters are ordered according to ν^t in such a way that, by definition, for each pair of letters a and b $\nu^t(a) \prec \nu^t(b)$ if $\lambda^t(a) \le \lambda^t(b)$. For example, if $A = \{a_1, a_2, a_3\}$ and $x_1 x_2 x_3 = a_3 a_2 a_3$, the possible orders can be as follows: $\nu^1 = (1, 2, 3)$, $\nu^2 = (3, 1, 2)$, $\nu^3 = (3, 2, 1)$, $\nu^4 = (3, 2, 1)$. In all other respects this method coincides with the book stack. (The set of all indexes $\{1, \ldots, S\}$ is divided into r subsets, etc.)

Obviously, after observing each letter x_t the value $\lambda^t(x_t)$ should be increased and the order ν^t should be changed. It is worth noting that there exist a data structure and algorithm which allow maintaining the alphabet letters ordered in such a way that the number of operations spent is constant, independent of the size of the alphabet. This data structure was described in [29, 38].

2.3.3 *Two-Faced Processes and the Choice of the Block Length for a Process Testing*

There are many methods for testing H_0 and H_1, where the bit stream is divided into words (blocks) of the length $s, s \geq 1$, and the sequence of the blocks $x_1 x_2 \dots x_s$, $x_{s+1} \dots x_{2s}, \dots$ is considered as letters, where each letter belongs to the alphabet $B_s = \{0, 1\}^s$ and has probability 2^{-s}, if H_0 is true. For instance, the two tests described above, methods from [39] and many other algorithms are of this type. That is why the questions of choosing the block length s will be considered here.

As was mentioned in the introduction, there exist two-faced processes, which, on the one hand, are far from being truly random, but, on the other hand, can be distinguished from truly random only in the case when the block length s is large. From the information theoretical point of view the two-faced processes can be simply described as follows. For a two-faced process, which generates letters from $\{0, 1\}$, the limit Shannon entropy is (much) less than 1 and, on the other hand, the s-order entropy (h_s) is maximal ($h_s = 1$ bit per letter) for relatively large s.

We describe two families of two-faced processes, $T(k, \pi)$ and $\bar{T}(k, \pi)$, where $k = 1, 2, \dots$ and $\pi \in (0, 1)$ are parameters. The processes $T(k, \pi)$ and $\bar{T}(k, \pi)$ are Markov chains of the connectivity (memory) k which generate letters from $\{0, 1\}$. It is convenient to define them inductively. The process $T(1, \pi)$ is defined by conditional probabilities $P_{T(1,\pi)}(0/0) = \pi, P_{T(1,\pi)}(0/1) = 1 - \pi$ (obviously, $P_{T(1,\pi)}(1/0) = 1 - \pi, P_{T(1,\pi)}(1/1) = \pi$). The process $\bar{T}(1, \pi)$ is defined by $P_{\bar{T}(1,\pi)}(0/0) = 1 - \pi, P_{\bar{T}(1,\pi)}(0/1) = \pi$. Assume that $T(k, \pi)$ and $\bar{T}(k, \pi)$ are defined and describe $T(k + 1, \pi)$ and $\bar{T}(k + 1, \pi)$ as follows:

$$P_{T(k+1,\pi)}(0/0u) = P_{T(k,\pi)}(0/u), P_{T(k+1,\pi)}(1/0u) = P_{T(k,\pi)}(1/u),$$
$$P_{T(k+1,\pi)}(0/1u) = P_{\bar{T}(k,\pi)}(0/u), P_{T(k+1,\pi)}(1/1u) = P_{\bar{T}(k,\pi)}(1/u),$$

and, vice versa,

$$P_{\bar{T}(k+1,\pi)}(0/0u) = P_{\bar{T}(k,\pi)}(0/u), P_{\bar{T}(k+1,\pi)}(1/0u) = P_{\bar{T}(k,\pi)}(1/u),$$
$$P_{\bar{T}(k+1,\pi)}(0/1u) = P_{T(k,\pi)}(0/u), P_{\bar{T}(k+1,\pi)}(1/1u) = P_{T(k,\pi)}(1/u)$$

for each $u \in B_k$ (here vu is a concatenation of the words v and u). For example,

$$P_{T(2,\pi)}(0/00) = \pi, P_{T(2,\pi)}(0/01) = 1 - \pi,$$
$$P_{T(2,\pi)}(0/10) = 1 - \pi, P_{T(2,\pi)}(0/11) = \pi.$$

The following theorem shows that the two-faced processes exist.

Theorem 2.3 *For each $\pi \in (0, 1)$ the s-order Shannon entropy (h_s) of the processes $T(k, \pi)$ and $\bar{T}(k, \pi)$ equals 1 bit per letter for $s = 0, 1, \ldots, k$ whereas the limit Shannon entropy (h_∞) equals $-(\pi \log_2 \pi + (1 - \pi) \log_2(1 - \pi))$.*

The proof of the theorem is given in Sect. 2.6.4, but here we consider the examples of "typical" sequences of the processes $T(1, \pi)$ and $\bar{T}(1, \pi)$ for π, say, 1/5. Such sequences could be as follows: 010101101010100101... and 0000111 110 0011 111 100 0.... We can see that each sequence contains approximately one half of 1's and one half of 0's. (That is why the first order Shannon entropy is 1 per letter.) On the other hand, the two sequences do not look truly random, because they, obviously, have too long subwords either 101010.. or 000..11111... (In other words, the second order Shannon entropy is much less than 1 per letter.) Hence, if a randomness test is based on estimation of frequencies of 0's and 1's only, then such a test will not be able to find deviations from randomness.

So, if we revert to the question about the block length of tests and take into account the existence of two-faced processes, it seems that the block length could be taken as large as possible. But it is not so. The following informal consideration could be useful for choosing the block length. The point is that statistical tests can be applied if words from the sequence

$$x_1x_2 \ldots x_s,\ x_{s+1} \ldots x_{2s},\ \ldots,\ x_{(m-1)s+1}x_{(m-1)s+2} \ldots x_{ms} \tag{2.8}$$

are repeated (at least a few times) with high probability (here ms is the sample length). Otherwise, if all words in (2.8) are unique (with high probability) when H_0 is true, it should be so for close hypotheses H_1. That is why a sensible test cannot be constructed basing on a division into s-letter words. So, the word length s should be chosen in such a way that some words from the sequence (2.8) are repeated with high probability when H_0 is true. So, now our problem can be formulated as follows. There is a binary sequence $x_1x_2 \ldots x_n$ generated by the Bernoulli source with $P(x_i = 0) = P(x_i = 1) = 1/2$ and we want to find such a block length s that the sequence (2.8) with $m = \lfloor n/s \rfloor$ contains some repetitions (with high probability). This problem is well known in probability theory and sometimes called the birthday problem. Namely, the standard statement of the problem is as follows. There are $S = 2^s$ cells and $m\,(= n/s)$ pellets. Each pellet is put in one of the cells with the probability $1/S$. It is known in Probability Theory that, if $m = c\sqrt{S}, c > 0$ then the average number of cells with at least two pellets equals $c^2\,(1/2 + o(1)\,)$, where S goes to ∞; see [22]. In our case the number of cells with at least two pellets is equal to the number of the words from the sequence (2.8) which are met two (or more) times. Taking into account that $S = 2^s, m = n/s$, we obtain from $m = c\sqrt{S}, c > 0$, an informal rule for choosing the length of words in (2.8):

$$n \asymp s2^{s/2} \tag{2.9}$$

where n is the length of a sample $x_1x_2 \ldots x_n$, and s is the block length. If s is much larger, the sequence (2.8) does not have repeated words (in case H_0) and it is difficult to build a sensible test. On the other hand, if s is much smaller, large classes of the alternative hypotheses cannot be tested (due to existence of the two-faced processes). It is worth noting that it is impossible to have a universal choice of s, because it is impossible to avoid the two-faced phenomenon. In other words this fact can be explained based on the following known result of Information Theory: it is impossible to have a guaranteed rate of code convergence universally for all ergodic sources; see [36]. That is why it is impossible to chose a universal length s. On the other hand, there are many applications where the word length s can be chosen experimentally. (But, of course, such experiments should be carried out with independent data.)

2.4 The Experiments

In this part we describe some experiments carried out to compare new tests with known ones. We will compare order test, book stack test, tests which are based on standard data compression methods, and tests from [33]. The point is that the tests from Rukhin and others are selected based on comprehensive theoretical and experimental analysis and can be considered as the state of the art in randomness testing. Besides, we will also test the method published in [39], because it was published later than the book [33].

We used data generated by the PRNG "RANDU" (described in [10]) and random bits from "The Marsaglia Random Number CDROM", see: http://stat.fsu.edu/diehard/cdrom/). RANDU is a linear congruent generator (LCG) which is defined by the following equality

$$X_{n+1} = (A\, X_n + C)\; mod\, M\,,$$

where X_n is nth generated number. RANDU is defined by parameters $A = 2^{16} + 3, C = 0, M = 2^{31}, X_0 = 1$. Those kinds of sources of random data were chosen because random bits from "The Marsaglia Random Number CDROM" are considered as good random numbers, whereas it is known that RANDU is not a good PRNG. It is known that the lowest digits of X_n are "less random" than the leading digits [19]; that is why in our experiments with RANDU we extract an eight-bit word from each generated X_i by formula $\hat{X}_i = \lfloor X_i/2^{23} \rfloor$.

The behavior of the tests was investigated for files of different lengths (see the tables below). We generated 100 different files of each length and applied each mentioned above test to each file with level of significance 0.01 (or less, see below). So, if a test is applied to a truly random bit sequence, on average one file from 100 should be rejected. All results are given in the tables, where integers in boxes are the number of rejected files (from 100). If a number of the rejections is not given for a certain length and test, it means that the test cannot be applied for files of such a length.

Table 2.1 Application to SPIHT

Name of test/length of file	50,000	100,000	500,000	1,000,000
Order test	56	100	100	100
Book stack	42	100	100	100
Parameters for both tests	s=20, $\|A_1\| = 5\sqrt{2^s}$			
RSS	4	75	100	100
Parameters	s=16	s=17	s=20	
RAR	0	0	100	100
ARJ	0	0	99	100
Frequency	2	1	1	2
Block frequency	1	2	1	1
Parameters	M=1000	M=2000	$M = 10^5$	M=20,000
Cumulative sums	2	1	2	1
Runs	0	2	1	1
Longest run of ones	0	1	0	0
Rank	0	1	1	0
Discrete Fourier transform	0	0	0	1
Nonoverlapping templates	–	–	–	2
Parameters				m=10
Overlapping templates	–	–	–	2
Parameters				m=10
Universal statistical	–	–	1	1
Parameters			L=6	L=7
			Q=640	Q=1280
Approximate entropy	1	2	2	7
Parameters	m=5	m=11	m=13	m=14
Random excursions	–	–	–	2
Random excursions variant	–	–	–	2
Serial	0	1	2	2
Parameters	m=6	m=14	m=16	m=8
Lempel-Ziv complexity	–	–	–	1
Linear complexity	–	–	–	3
Parameters				M=2500

Table 2.1 contains information about testing of sequences of different lengths generated by RANDU, whereas Table 2.2 contains results of application of all tests to 5,000,000-bit sequences either generated by RANDU or taken from "The Marsaglia Random Number CDROM". For example, the first number of the second line of Table 2.1 is 56. It means that there were 100 files of the length 50,000 bits generated by PRNG RANDU. When the Order test was applied, the hypothesis H_0 was rejected 56 times from 100 (and, correspondingly, H_0 was accepted 44 times). The first number of the third line shows that H_0 was rejected 42 times when the Book Stack test was applied to the same 100 files. The third number of the second

Table 2.2 Results of application of all tests to 5,000,000-bit sequences either generated by RANDU or taken from The Marsaglia Random Number CDROM

Name of test/length of file	5,000,000 (*RANDU*)	5,000,000 (*random bits*)
Order test	100	3
Book stack	100	0
Parameters for both tests	s=24, $\|A_1\| = 5\sqrt{2^s}$	
RSS	100	1
Parameters	s=24	s=24
RAR	100	0
ARJ	100	0
Frequency	2	1
Block frequency	2	1
Parameters	$M = 10^6$	$M = 10^5$
Cumulative sums	3	2
Runs	2	2
Longest run of ones	2	0
Rank	1	1
Discrete Fourier transform	89	9
Nonoverlapping templates	5	5
Parameters	m=10	m=10
Overlapping templates	4	1
Parameters	m=10	m=10
Universal statistical	1	2
Parameters	L=9	L=9
	Q=5120	Q=5120
Approximate entropy	100	89
Parameters	m=17	m=17
Random excursions	4	3
Random excursions variant	3	3
Serial	100	2
Parameters	m=19	m=19
Lempel-Ziv complexity	0	0
Linear complexity	4	3
Parameters	M=5000	M=2500

line shows that the hypothesis H_0 was rejected 100 times when the Order test was applied for testing of 100 100,000-bit files generated by RANDU, etc.

Let us first give some comments about the tests, which are based on popular data compression methods RAR and ARJ. In those cases we applied each method to a file and first estimated the length of compressed data. Then we use the test $\Gamma_{\alpha,\varphi}^{(n)}$ with the critical value (2.1) as follows. The alphabet size $|A| = 2^8 = 256$, $n \log |A|$ is simply the length of file (in bits) before compression (whereas n is the length in bytes). So, taking $\alpha = 0.01$, from (2.1), we see that the hypothesis about randomness (H_0)

should be rejected if the length of compressed file is less than or equal to $n \log |A| - 8$ bits. (Strictly speaking, in this case $\alpha \le 2^{-7} = 1/128$.) So, taking into account that the length of computer files is measured in bytes, this rule is very simple: if the n-byte file is really compressed (i.e. the length of the encoded file is $n - 1$ bytes or less), this file is not random (and H_0 is rejected). So, the following tables contain numbers of cases where files were really compressed.

Let us now give some comments about parameters of the considered methods. As was mentioned, we investigated all methods from the book [33], the test from [39] (RSS test for short), the two tests described above based on data compression algorithms, the order tests and the book stack test. For some tests there are parameters, which should be specified. In such cases the values of parameters are given in the table in the row which follows the test results. There are some tests from the book [33] where parameters can be chosen from a certain interval. In such cases we repeated all calculations three times, taking the minimal possible value of the parameter, the maximal one and the average one. Then the data for the case when the number of rejections of the hypothesis H_0 is maximal is given in the table.

The choice of parameters for RSS, the book stack test and the order test was made on the basis of special experiments, which were carried out for independent data. (Those algorithms are implemented as a Java program and can be found on the internet, see http://web.ict.nsc.ru/~rng/.) In all cases such experiments have shown that for all three algorithms the optimal blocklength is close to the one defined by informal equality (2.9).

We can see from the tables that the new tests can detect non-randomness more efficiently than the known ones. Seemingly, the main reason is that RSS, book stack tests and order test deal with as large blocklengths as possible, whereas many other tests are focused on other goals. The second reason could be an ability for adaptation. The point is that the new tests can find subwords which are more frequent than others and use them for testing, whereas many other tests are looking for particular deviations from randomness.

In conclusion, we can say that the obtained results show that the new tests, as well as the ideas of Information Theory in general, can be useful tools for randomness testing.

2.5 Analysis of Stream Ciphers and Random Number Generators Used in Cryptography

As an example of application of the suggested approach we show how the book stack test was applied for checking a real stream cipher (ZK-Crypt) and a real PRNG (RC4) recommended for cryptography systems. The results described below are mainly based on the papers [8, 9, 25].

The stream cipher ZK-Crypt is a submission to the ECRYPT stream cipher project. The book stack test for randomness was applied to this cipher. It is experimentally shown that the keystream generated from ZK-Crypt can be distinguished from random with about 2^{25} output bits.

2.5.1 *The Distinguishing Attack on ZK-Crypt Cipher*

ZK-Crypt [16] is a stream cipher proposed as a candidate to the ECRYPT Stream Cipher Project. It is required that a keystream of any stream cipher should be truly random, i.e., by definition, a generated sequence could be interpreted as the result of the flips of a "fair" coin with sides that are labelled "0" and "1" (for short, it is called a random sequence).

In this part, we describe some experiments intended to detect non-randomness of the ZK-Crypt keystream. In cryptography such an attempt is called a distinguishing attack. In other words, we describe a distinguishing attack on ZK-Crypt cipher, whose goal is to distinguish the keystream of the cipher from the random sequence. For this purpose the book stack test was applied to keystream sequences generated by the ZK-Crypt cipher. It turns out that the keystream sequences are far from random when their length is about 2^{25} bits.

The behavior of the ZK-crypt keystream was investigated for 100 randomly chosen keys by the book stack. When the test was applied, the keystream sequence was divided into 32-bit words (blocks) and any block was considered as a letter from an alphabet whose size is 2^{32}. According to the test, any alphabet letter (i.e. a 32-bit word) was assigned an index and the set of all indexes was divided into two subsets $A_1 = \{1, \ldots, 2^{16}\}$ and $A_2 = \{2^{16} + 1, \ldots, 2^{32}\}$.

We generated files of different lengths for each key (see the tables below) and applied the book stack test to each file with level of significance 0.001. So, if a test is applied to a random bit sequence, on average one file from 1000 should be recognized as non-random. All results are given in the table, where integers in the cells are the numbers of files recognized as non-random (out of 100).

Having taken into account that, on average, only 0.1 files from 100 should be recognized as non-random if sequences are random, we see that the keystream sequences are far from random when their length is about 2^{25}.

2.5.2 *Analysis of the PRNG RC4*

The RC4 has been one of the most popular stream ciphers since it was proposed by Ron Rivest in 1987. It is required that a keystream of RC4 (and any stream cipher) should be truly random. There are some papers where it is shown theoretically that the RC4 keystream is not truly random [7, 12, 13, 15, 30] and a paper [6, 14] where non-randomness of RC4 was experimentally verified with over 56 trillion samples ($>2^{55}$ bits) of RC4 output.

Table 2.3 Number of files generated by ZK-Crypt and recognized as non-random (from 100)

Length (bits)	$16777216 = 2^{24}$	$67108864 = 2^{26}$	$268435456 = 2^{28}$	$1073741824 = 2^{30}$
Non-random	25	51	97	100

Table 2.4 Number of files generated by RC4 and recognized as non-random (from 100)

Length (bits)	2^{31}	2^{32}	2^{33}	2^{34}	2^{33}	2^{36}	2^{37}	2^{38}	2^{39}
Non-random	6	12	17	23	37	74	95	99	100

In this paper, we describe some experiments intended to detect non-randomness of the RC4 keystream. For this purpose the book stack test was applied to keystream sequences generated by RC4. It turns out that the keystream sequences are far from random when their length is about 2^{32} bits. We consider the most popular version of RC4, in which the number of bits in a single output is eight.

The behavior of the RC4 keystream was investigated for 100 randomly chosen keys by the above described book stack test. To apply the test we divided the keystream sequence into 16-bit words (blocks) and each block was considered as a letter from the alphabet of size 2^{16}. According to the test, any alphabet letter (i.e., a 16-bit word) was assigned an index and the set of all indexes was divided into two subsets, $A_1 = \{1, \ldots, 16\}$ and $A_2 = \{17, \ldots, 2^{16}\}$.

We generated files of different lengths for each randomly chosen 256-bit key (see Tables 2.3 and 2.4) and applied the book stack test to each file with level of significance 0.05. So, if a test is applied to a random bit sequence, on average five files from 100 should be recognized as non-random. All results are given in the table, where integers in the cells are the numbers of files recognized as non-random (out of 100).

Having taken into account that, on average, only five files from 100 should be recognized as non-random if sequences are random we see that the keystream sequences are far from random when their length is about 2^{32}.

2.6 A Statistical Attack on Block Ciphers

2.6.1 Cryptanalysis of Block Ciphers

Cryptanalysis of block ciphers attracts much research, and new results in the field are always beneficial for improving constructions of the ciphers. Sometimes the complexity of a new attack (measured in the number of memory units and operations required for mounting such an attack) might be quite large. Nevertheless, even if a relatively small decrease in the attack complexity is achieved, in comparison with previously known methods, this can motivate further development of the cipher design. Thus linear cryptanalysis of DES (see [28]) requires 2^{43} known plaintext-ciphertext pairs and is generally considered infeasible in practice. But it has made an important impact on the design principles of modern block ciphers, which are now resistant to this kind of attack. In this contribution, we suggest a new attack for block ciphers, referred to as a "gradient statistical attack". We show the ability of mounting this attack with respect to the ciphers, for which no attacks are known better than the exhaustive key search.

Consider a (symmetric-key) block cipher with blocklength n, key length s, and encryption function $E(x, K)$, where $x \in \{0, 1\}^n$ denotes a plaintext block, and $K \in \{0, 1\}^s$ a secret key. The typical values of n and s for modern block ciphers are $n = 64$ or $n = 128$, $s = 128$ bits. The majority of block ciphers are iterated, i.e., involve many rounds of transformations usually bracketed by some prologue and epilogue. Either of these, in turn, can sometimes be divided into a number of more simple steps. In respect of this iterated structure, the secret key K is expanded into a sequence of subkeys (or round keys) $k_1, k_2, \ldots, k_t$, where t is the number of "simple steps" in a block cipher. Denote by x_0 the initial state of block x, and by x_i the state after the ith step. So the complete encryption is $x_t = E(x_0, K)$ and this can be written as

$$x_1 = E_1(x_0, k_1), \quad x_2 = E_2(x_1, k_2), \quad \ldots, \quad x_t = E_t(x_{t-1}, k_t), \tag{2.10}$$

where E_i denotes the encrypting transformation at the ith step.

Let us give an example. Consider the cipher RC5 [31] with the blocklength 64 and number of rounds r. The encrypting process, with reference to (2.10), is as follows:

$$\begin{array}{lll}
\text{input:} & x_0 = (a, b) & \\
\text{prologue:} & a \leftarrow a + k_1 & x_1 = E_1(x_0, k_1) \\
 & b \leftarrow b + k_2 & x_2 = E_2(x_1, k_2) \\
\text{round 1:} & a \leftarrow ((a \oplus b) \hookleftarrow b) + k_3 & x_3 = E_3(x_2, k_3) \\
 & b \leftarrow ((b \oplus a) \hookleftarrow a) + k_4 & x_4 = E_4(x_3, k_4) \\
 & \ldots & \\
\text{round r:} & a \leftarrow ((a \oplus b) \hookleftarrow b) + k_{2r+1} & x_{t-1} = E_{t-1}(x_{t-2}, k_{t-1}) \\
 & b \leftarrow ((b \oplus a) \hookleftarrow a) + k_{2r+2} & x_t = E_t(x_{t-1}, k_t) \\
\text{output:} & x_t = (a, b) & (t = 2r + 2)
\end{array}$$

The length of each subkey is 32 bits. Many other ciphers, including RC6 and AES, may also be represented by (2.10) with relatively small, e.g., 32 bits or less, subkeys.

We suggest a chosen-plaintext attack for the cipher which can be represented by (2.10) with relatively small subkeys. Denote the lengths of the secret key and each subkey by $|K|$ and $|k|$, respectively. The exhaustive key search requires $O(2^{|K|})$ operations (decrypt with $K = 0, 1, \ldots$ until a known x is obtained). Meanwhile, the suggested attack requires $O(mt2^{|k|})$ operations, where m is the number of ciphertext blocks sufficient for statistical analysis. The attack succeeds in finding the correct subkeys (instead of K itself), provided the statistical test is able to detect deviations from randomness in a sequence of m blocks.

The experimental part of our research was first concentrated on the RC5 block cipher. Currently, we came to the point that RC5 with eight rounds can be broken by our attack with 2^{33} chosen plaintext-ciphertext pairs. To our knowledge, this is the best result to date (cf. [17, 40]). The data obtained allow to conclude that the suggested attack can also be applied to other ciphers of the form (2.10), e.g., RC6 and AES.

2.6.2 Description of the Attack

The suggested attack belongs to a class of chosen-plaintext attacks. Upon that kind of attack, a cryptanalyst is able to input any information to the cipher and observe the corresponding output. Her aim is to recover the secret key or, what is almost the same, the round keys. Such attacks are of practical interest and it is assumed that the block ciphers must be secure against them.

We consider the block ciphers that can be described by (2.10) (this includes many of the modern ciphers, if not all). Notice that the sequence for decryption corresponding to (2.10) is

$$x_{t-1} = D_t(x_t, k_t), \quad x_{t-2} = D_{t-1}(x_{t-1}, k_{t-1}), \quad \dots, \quad x_0 = D_1(x_1, k_1), \qquad (2.11)$$

where D_i denotes the decrypting transformation inverse to E_i.

One of the requirements of a block cipher is that, given a sequence of different blocks as input, the cipher must output a sequence of bits that looks random. We shall loosely call bit sequences "more random" or "less random" depending on how much they differ from a truly random sequence. One way to measure randomness is to use some statistic on the sequence with the property that less random sequences have greater statistics (to within some probability of error in decision). This may be a well-known x^2 statistic subjected to χ^2 distribution. Denote such a statistic by $\gamma(x)$, where x is a bit sequence.

Denote by $\alpha_1, \alpha_2, \dots, \alpha_m$ a sequence of input blocks. Let all the blocks be non-random and pairwise different. A possible example would be $\alpha_1 = 1, \alpha_2 = 2, \dots,$ $\alpha_m = m$, where the numbers are written as n-bit words. For a good block cipher, the encrypted sequence

$$E(\alpha_1, K), \quad E(\alpha_2, K), \quad \dots, \quad E(\alpha_m, K)$$

must look random, for any K. Now recall (2.10). Apply only one step of encryption to the input sequence, denoting the result by $\beta_1, \beta_2, \dots, \beta_m$:

$$\beta_1 = E_1(\alpha_1, k_1), \quad \beta_2 = E_1(\alpha_2, k_1), \quad \dots, \quad \beta_m = E_1(\alpha_m, k_1).$$

We claim that the sequence β is more random than the sequence α, i.e., $\gamma(\beta) < \gamma(\alpha)$. After the second step of encryption, the sequence

$$E_2(\beta_1, k_2), \quad E_2(\beta_2, k_2), \quad \ldots, \quad E_2(\beta_m, k_2)$$

is more random than β, and so on. Each step of encryption increases the degree of randomness.

Notice the obvious consequence: in decryption according to (2.11), the randomness of the data decreases from step to step. For example, the sequence

$$D_1(\beta_1, k_1), \quad D_1(\beta_2, k_1), \quad \ldots, \quad D_1(\beta_m, k_1),$$

which is α, is less random than β. But what is important is that this is true only if the decryption is done with the valid key. If the key is not valid, denote it by k_1'; then the sequence

$$\alpha_1' = D_1(\beta_1, k_1'), \quad \alpha_2' = D_1(\beta_2, k_1'), \quad \ldots, \quad \alpha_m' = D_1(\beta_m, k_1')$$

will be *more* random than β, $\gamma(\alpha') < \gamma(\beta)$. This is because decrypting with a different key corresponds to further encrypting with that key, which is the well-known multiple encryption principle. So, generally, decryption with an invalid round key increases randomness, while decryption with the valid round key decreases randomness. This difference can be detected by a statistical test.

The suggested general statistical attack is mounted as follows. First encrypt the sequence $\alpha_1, \alpha_2, \ldots, \alpha_m$, defined above. Denote the output sequence by ω,

$$\omega_1 = E(\alpha_1, K), \quad \omega_2 = E(\alpha_2, K), \quad \ldots, \quad \omega_m = E(\alpha_m, K).$$

(Recall that the cipher involves t rounds or steps, and the length of subkey at each step is $|k|$.)

Now begin the main procedure of key search. For all $u \in \{0, 1\}^{|k|}$ compute a sequence

$$\Gamma_t(u) = D_t(\omega_1, u), \quad D_t(\omega_2, u), \quad \ldots, \quad D_t(\omega_m, u)$$

and estimate its randomness, i.e., compute $\gamma(\Gamma_t(u))$. Find such u^* for which $\gamma(\Gamma_t(u^*))$ is maximal. Assume that unknown subkey $k_t = u^*$. Note that the number of operations at this stage is $O(m2^{|k|})$.

After that, based upon the sequence $\Gamma_t(k_t)$, repeat the similar computations to find subkey k_{t-1}. Using $\Gamma_{t-1}(k_{t-1})$ find k_{t-2}, and so on down to k_1. The total number of operations to recover all subkeys is $O(mt2^{|k|})$.

2.6.3 Variants and Optimizations

We have described the idea of the attack in a pure form. In this section, we discuss some implementation variants.

1. The measure of randomness γ is a parameter. One may apply different measures not only for different ciphers, but also for different rounds of the same cipher. As stated above, any statistical test applicable to checking for the main hypothesis H_0 that the binary sequence is truly random versus the opposite hypothesis H_1 that the sequence is not may be used for that purpose. There are nowadays many works devoted to statistical tests due to their great importance for various applications. Thus the US National Institute of Standards and Technology (NIST) has done research of known tests for randomness and recommended 16 tests for use in cryptography, see [33]. However, as shown above, the tests which are based on ideas of information theory outperform all the tests from [33].

   The test RSS from [39] runs in $O(m)$ time on a sample of length m. The book stack test runs in $O(m \log m)$ time, being slightly more powerful. These tests are at the heart of the proposed attack.
2. The sequence length m was chosen to be constant for simplicity of description. In fact, this length may vary in each round. More lengthy sequences are needed when cipher output becomes more random, i.e., in the last rounds. The value of m depends on the power of the statistical test and usually can be determined experimentally (see the next section).
3. Some division of the encryption process into simple steps may result in a situation where, at a particular step, only part of the block is affected. This may reduce the effective length of the sequence to be tested. Thus the division of RC5 into steps as shown in the example assumes that at the odd steps $(1, 3, \ldots, t-1)$ only the left half of the block, a, changes, and at the even $(2, 4, \ldots, t)$ the right half, b. Therefore only one half of the sequence, virtually $m/2$ blocks, is to be subjected to statistical tests.
4. When searching for a relevant subkey, it is reasonable to keep not a single but several candidate subkeys, say, s different values of $u \in \{0,1\}^{|k|}$ for which $\gamma(\Gamma_j(u))$ is maximal. Besides, when searching for simple sequences and subkeys, the sequential methods, analogous to sequential criteria in statistics, are appropriate.
5. The initial non-random sequence $\alpha_1, \alpha_2, \ldots, \alpha_m$ may be constructed in different ways. For example, it seems to make sense to choose the sequence in which consecutive words (α_i, α_{i+1}) not only contain many equal bits but differ in at most one bit (so called Gray code). Other choices may reflect the peculiarities of a particular cipher.
6. The last modification may be connected with the fact that many modern ciphers with great number of rounds convert even absolutely non-random input sequence into something "quite random" (at least not distinguishable from a truly random by known tests in acceptable time). Let, for instance, the cipher have r rounds

and for some simple initial sequence $\alpha_1^0, \alpha_2^0, \ldots, \alpha_m^0$ the sequences

$$\begin{aligned}
\alpha^1 &= E_1(\alpha_1^0, k_1), \quad E_1(\alpha_2^0, k_1), \quad \ldots, \quad E_1(\alpha_m^0, k_1), \\
\alpha^2 &= E_2(\alpha_1^1, k_2), \quad E_2(\alpha_2^1, k_2), \quad \ldots, \quad E_2(\alpha_m^1, k_2), \\
&\ldots \\
\alpha^d &= E_d(\alpha_1^{d-1}, k_d), \quad E_d(\alpha_2^{d-1}, k_d), \quad \ldots, \quad E_d(\alpha_m^{d-1}, k_d)
\end{aligned}$$

not be random under all subkeys $k_1, \ldots, k_d$, $d < r$. Then the suggested attack may be modified in the following manner. For each set of subkeys $k_{d+1}, \ldots, k_r$ of rounds $d+1, \ldots, r$ apply the procedure described in Sect. 2.2 and find unknown subkeys $k_1, \ldots, k_d$. In other words, find subkeys $k_{d+1}, \ldots, k_r$ by exhaustive search, and $k_1, \ldots, k_d$ by the method of Sect. 2.2. To maintain this combined attack $O(m2^{(r-d)|k|}d2^{|k|})$ operations are required, which, depending on certain parameters, can be smaller than $O(2^{|K|})$ for exhaustive key search.

2.6.4 Experiments with RC5

Our experiments were planned as follows. First, we analysed the degree of randomness of encrypted sequences as a function of the number of steps of encryption. The goal was to find the maximal number of steps at which the tests can tell the encrypted sequence from truly random. Second, we examined the claim our attack is based upon, namely, whether decrypting with a wrong subkey increases randomness in comparison with decrypting using a valid subkey; more exactly, whether these are distinguishable by the tests. Third, based on the results obtained, we started the suggested attack (now in progress). The experiments were carried out on a multiprocessor system, which contains ten 1 GHz Alpha processors with 1 GB of memory per device, of which only five devices were available for computations. Programs were written in the C language and run under the Linux operating system. RSS from [39] was chosen as a method of statistical analysis.

To test the statistical properties of RC5 up to six steps (according to the example in Chap. 1), 100 sequences

$$\alpha_1\alpha_2\ldots\alpha_m, \quad \alpha_{m+1}\alpha_{m+2}\ldots\alpha_{2m}, \quad \ldots, \quad \alpha_{99m+1}\alpha_{99m+2}\ldots\alpha_{100m},$$

where $\alpha_i = i$ in n-bit binary representation, were encrypted under 10 randomly selected keys $K_1, \ldots, K_{10}$. Table 2.5 shows the number of cases in which the output sequence of the cipher was declared non-random (with significance level 0.0001). We can see that the cipher output is definitely non-random since, otherwise, it would have been rejected only $0.0001 \times 100 = 0.1$ times out of 100.

Computations for a greater number of steps were carried out with one sequence $\alpha_1\alpha_2\ldots\alpha_m$, m substantially increased, under a number of randomly selected keys

Table 2.5 The number of sequences (out of 100) declared non-random

Key	$t = 2$ $m = 2^{18}$	$t = 3$ $m = 2^{18}$	$t = 4$ $m = 2^{18}$	$t = 5$ $m = 2^{20}$	$t = 6$ $m = 2^{20}$
K_1	100	63	64	51	52
K_2	100	100	100	74	70
K_3	100	61	61	17	17
K_4	100	81	78	62	64
K_5	100	100	100	65	6
K_6	100	85	86	12	9
K_7	100	100	100	11	8
K_8	100	98	99	99	99
K_9	100	80	79	14	14
K_{10}	100	100	100	7	5

Table 2.6 The number of sequences declared non-random

t	m	Number of keys	Number of non-random outputs
10	2^{28}	30	30
11	2^{29}	22	10
12	2^{31}	6	6
13	2^{32}	6	6
14	2^{32}	6	5
15	2^{33}	3	3

Table 2.7 The number of sequences declared non-random

	Valid subkey	u_1	u_2	u_3	u_4	u_5
$t = 8$	10	4	4	4	3	3
$t = 9$	5	0	0	0	0	0

(Table 2.6). We can see that the encrypted sequence is stably distinguished from the truly random ones up to the fifteenth step (with significance level 0.01), which corresponds to the eighth round of RC5.

To examine the distinguishability of the sequences decrypted with a valid and invalid subkeys, we used the initial sequence of length $m = 2^{24}$ encrypted with eight and nine steps under a randomly selected key K and corresponding subkeys k_8 and k_9. In each case the output sequence was decrypted one step downwards with the valid and five randomly chosen invalid subkeys $u_1, \ldots, u_5$. All computations were repeated ten times. Table 2.7 shows the number of cases in which the sequence was declared non-random. One can see that the situations of decrypting using valid and invalid keys are distinguished reliably (ten against four and five against zero).

The experiments presented confirm our assumptions as to principal possibility of the suggested gradient statistical attack. First, the encrypted sequence gets more random as the number of rounds increases. And second, a sequence decrypted with invalid subkey is more random than one decrypted with valid subkey, and the test can detect this.

Appendix

Proof of Theorem 2.1 First we estimate the number of words $\varphi_n(u)$ whose length is less than or equal to an integer τ. Obviously, at most one word can be encoded by the empty codeword, at most two words by the words of the length 1, ..., at most 2^i by the words of length i, etc. Having taken into account that the codewords $\varphi_n(u) \neq \varphi_n(v)$ for different u and v, we obtain the inequality

$$|\{u : |\varphi_n(u)| \leq \tau\}| \leq \sum_{i=0}^{\tau} 2^i = 2^{\tau+1} - 1.$$

From this inequality and (2.1) we can see that the number of words from the set $\{A^n\}$, whose codelength is less than or equal to $t_\alpha = n \log|A| - \log(1/\alpha) - 1$, is not greater than $2^{n \log|A| - \log(1/\alpha)}$. So, we obtained that

$$|\{u : |\varphi_n(u)| \leq t_\alpha\}| \leq \alpha |A|^n.$$

Taking into account that all words from A^n have equal probabilities if H_0 is true, we obtain from the last inequality, (2.1) and the description of the test $\Gamma_{\alpha,\varphi}^{(n)}$ that

$$Pr\{|\varphi_n(u)| \leq t_\alpha|\} \leq (\alpha|A|^n/|A|^n) = \alpha$$

if H_0 is true. The theorem is proved.

Proof of Claim 2.1 The proof is based on a direct calculation of the probability of rejection for a case where H_0 is true. From the description of the test $\Upsilon_{\alpha,\varphi}^{(n)}$ and definition of g (see (2.8)) we obtain the following chain of equalities.

$$Pr\{H_0 \textit{ is rejected}\} = Pr\{\,|\varphi_n(u)| = g\}$$

$$+ Pr\{|\varphi_n(u)| = g+1\}\,(\,1 \;-\; (\sum_{j=1}^{g+1} |A_j| \;-\; \alpha|A|^n\,)/|A_{g+1}|\,)\,)$$

$$= \frac{1}{A^n}\,(\sum_{j=0}^{g} |A_j| \;+\; |A_{g+1}|\,(\,1 \;-\; (\sum_{j=1}^{g+1} |A_j| \;-\; \alpha|A|^n\,)/|A_{g+1}|\,)\,) = \alpha.$$

The claim is proved.

Proof of Claim 2.2 We can think that $\hat{t}_\alpha$ in (2.4) is an integer. (Otherwise, we obtain the same test taking $\lfloor \hat{t}_\alpha \rfloor$ as a new critical value of the test.) From the Kraft inequality (2.3) we obtain that

$$1 \geq \sum_{u \in A^n} 2^{-|\varphi_n(u)|} \geq |\{u : |\,\varphi_n(u)| \leq \hat{t}_\alpha\}|\; 2^{-\hat{t}_\alpha}.$$

This inequality and (2.4) yield:

$$|\{u : |\varphi_n(u)| \leq \hat{t}_\alpha\}| \leq \alpha |A|^n.$$

If H_0 is true then the probability of each $u \in A^n$ equals $|A|^{-n}$ and from the last inequality we obtain that

$$Pr\{|\varphi(u)| \leq \hat{t}_\alpha\} = |A|^{-n} \, |\{u : |\varphi_n(u)| \leq \hat{t}_\alpha\}| \leq \alpha$$

if H_0 is true. The claim is proved.

Proof of Theorem 2.3 We prove the theorem for the process $T(k, \pi)$, but this proof is valid for $\bar{T}(k, \pi)$, too. First we show that

$$p^*(x_1 \dots x_k) = 2^{-k}, \tag{2.12}$$

$(x_1 \dots x_k) \in \{0, 1\}^k$, is a stationary distribution for the processes $T(k, \pi)$ (and $\bar{T}(k, \pi)$) for all $k = 1, 2, \dots$ and $\pi \in (0, 1)$. For any values of $k, k \geq 1$, (2.4) will be proved if we show that the system of equations

$$\begin{aligned} P_{T(k,\pi)}(x_1 \dots x_d) &= P_{T(k,\pi)}(0x_1 \dots x_{d-1}) \, P_{T(k,\pi)}(x_d/0x_1 \dots x_{d-1}) \\ &\quad + P_{T(k,\pi)}(1x_1 \dots x_{d-1}) \, P_{T(k,\pi)}(x_d/1x_1 \dots x_{d-1}) \end{aligned}$$

has the solution $p(x_1 \dots x_d) = 2^{-d}$, $(x_1 \dots x_d) \in \{0, 1\}^d$, $d = 1, 2, \dots, k$. It can be easily seen for $d = k$, if we take into account that, by definition of $T(k, \pi)$ and $\bar{T}(k, \pi)$, the equality $P_{T(k,\pi)}(x_k/0x_1 \dots x_{k-1}) + P_{T(k,\pi)}(x_k/1x_1 \dots x_{k-1}) = 1$ is valid for all $(x_1 \dots x_k) \in \{0, 1\}^k$. From this equality and the low of total probability we immediately obtain (2.4) for $d < k$.

Let us prove the second claim of the theorem. From the definition $T(k, \pi)$ and $\bar{T}(k, \pi)$ we can see that either $P_{T(k,\pi)}(0/x_1 \dots x_k) = \pi$, $P_{T(k,\pi)}(1/x_1 \dots x_k) = 1 - \pi$ or $P_{T(k,\pi)}(0/x_1 \dots x_k) = 1 - \pi$, $P_{T(k,\pi)}(1/x_1 \dots x_k) = \pi$. That is why $h(x_{k+1}/x_1 \dots x_k) = -(\pi \log_2 \pi + (1 - \pi) \log_2(1 - \pi))$ and, hence, $h_\infty = -(\pi \log_2 \pi + (1 - \pi) \log_2(1 - \pi))$. The theorem is proved.

References

1. Aho, A.V., Hopcroft, J.E., Ullman, J.D.: The Design and Analysis of Computer Algorithms. Addison-Wesley, Reading (1976)
2. Bently, J.L., Sleator, D.D., Tarjan, R.E., Wei, V.K.: A locally adaptive data compression scheme. Commun. ACM **29**, 320–330 (1986)
3. Billingsley, P.: Ergodic Theory and Information Wiley, New York (1965)
4. Calude, C.S.: Information and Randomness - An Algorithmic Perspective, 2nd edn. (Revised and Extended). Springer, Berlin (2002)

5. Cover, T.M., Thomas, J.A.: Elements of Information Theory. Wiley-Interscience, New York (2006)
6. Crowley, P.: Small bias in RC4 experimentally verified. http://www.ciphergoth.org/crypto/rc4/ (2002)
7. Dawson, E., Gustafson, H., Henricksen, M., Millan, B.: Evaluation of RC4 Stream Cipher . http://www.ipa.go.jp/security/enc/CRYPTREC/fy15/ (2002)
8. Doroshenko, S., Fionov, A., Lubkin, A., Monarev, V., Ryabko, B.: On ZK-Crypt, Book Stack, and Statistical Tests. Cryptology ePrint Archive: Report 2006/196. http://eprint.iacr.org/2006/196.pdf (2006)
9. Doroshenko, S., Ryabko, B.: The experimental distinguishing attack on RC4. Cryptology ePrint Archive: Report 2006/070. http://eprint.iacr.org/2006/070 (2006)
10. Dudewicz, E.J., Ralley, T.G.: The Handbook of Random Number Generation and Testing with TESTRAND Computer Code. American Sciences Press, Columbus (1981)
11. Elias, P.: Interval and recency rank source coding: two on-line adaptive variable-length schemes. IEEE Trans. Inf. Theory **3**(1) , 3–10 (1987)
12. Fluhrer, S., McGrew, D.: Statistical analysis of the alleged RC4 keystream generator source. In: Proceedings of the 7th International Workshop on Fast Software Encryption Table of Contents. Lecture Notes in Computer Science, vol. 1978, pp. 19–30 (2000)
13. Gallager, R.G.: Information Theory and Reliable Communication. Wiley, New York (1968)
14. Gilbert, H., Handschuh, H., Joux, A., Vaudenay, I.S.: A statistical attack on RC6. In: Fast Software Encryption. Lecture Notes in Computer Science, vol. 1978, pp. 64–74 (2001)
15. Golic, J. D.: Iterative probabilistic cryptanalysis of RC4 keystream generator. In: Australasian Conference on Information Security and Privacy (ACISP), pp. 220–233 (2000)
16. Gressel, C., Granot, R., Vago, G.: ZK-Crypt. ECRYPT Stream Cipher Project Report 2005. Available at http://www.ecrypt.eu.org/stream/zkcrypt.html (2005)
17. Kaliski, B.S., Jr., Yin, Y.L.: On differential and linear cryptanalysis of the RC5 encryption algorithm. In: Advances in Cryptology — CRYPTO'95. Lecture Notes in Computer Science, vol. 963, pp. 171–184 (1995)
18. Kendall, M.G., Stuart, A.: The Advanced Theory of Statistical. Inference and Relationship, vol. 2. London (1961)
19. Knuth, D.E.: The Art of Computer Programming, vol. 2. Addison-Wesley, Reading (1981)
20. Kolmogorov, A.N.: Three approaches to the quantitative definition of information. Probl. Inf. Transm. **1**, 3–11 (1965)
21. Knudsen, R.L., Meier, W.: Correlation in RC6. Private communication. http://www.ii.uib.no/larsr/papers/rc6.ps
22. Kolchin, V.F., Sevastyanov, B.A., Chistyakov, V.P.: Random Allocations. Halsted Press, New York (1978)
23. L'Ecuyer, P.: Uniform random number generation. Ann. Oper. Res. **53**(1), 77–120 (1994)
24. Li, M., Vitanyi, P.: An Introduction to Kolmogorov Complexity and Its Applications, 2nd edn. Springer, New York (1997)
25. Lubkin, A., Ryabko, B.: The distinguishing attack on ZK - Crypt cipher. eSTREAM, ECRYPT Stream Cipher Project, 2005, Report 2005/076. http://www.ecrypt.eu.org/stream (2005)
26. Marsaglia, G.: The structure of linear congruential sequences. In: Zaremba, S.K. (ed.) Applications of Number Theory to Numerical Analysis, pp. 248–285. Academic, New York (1972)
27. Marsaglia, G., Zaman, A.: Monkey tests for random number generators. Comput. Math. Appl. **26**, 1–10 (1993)
28. Menzes, A., van Oorschot, P., Vanstone, S.: Handbook of Applied Cryptography. CRC Press, Boca Raton (1996)
29. Moffat, A.: An improved data structure for cumulative probability tables. Softw. Pract. Exp. **29**(7), 647–659 (1999)
30. Pudovkina, M.: Statistical weaknesses in the alleged RC4 keystream generator. Cryptology ePrint Archive: Report 2002/171. http://eprint.iacr.org/2002/171 (2002)

31. Rivest, R.L.: The RC5 encryption algorithm. In: Preneel, B. (ed.) Fast Software Encryption. Second International Workshop. Lecture Notes in Computer Science, vol. 1008, pp. 86–96. Springer, Berlin (1995)
32. Rozanov, Yu.A.: The Random Processes. Nauka, Moscow (1971)
33. Rukhin, A., Soto, A.J., Nechvatal, J., Smid, M., Barker, E., Leigh, S., Levenson, M., Vangel, M., Banks, D., Heckert, A., Dray, J., Vo, S.: A statistical test suite for random and pseudorandom number generators for cryptographic applications. NIST Special Publication 800–22 (with revision dated April 2010). http://csrc.nist.gov/publications/nistpubs/800-22-rev1a/SP800-22rev1a.pdf (2010)
34. Ryabko, B.Ya.: Information compression by a book stack. Probl. Inf. Transm. **16**(4), 16–21 (1980)
35. Ryabko, B.Ya.: A locally adaptive data compression scheme. Commun. ACM **30**(9), 792 (1987)
36. Ryabko, B.Ya.: Prediction of random sequences and universal coding. Probl. Inf. Transm. **24**(2), 87–96 (1988)
37. Ryabko, B., Fionov, A.: Basics of Contemporary Cryptography for IT Practitioners. World Scientific, Singapore (2005)
38. Ryabko, B., Rissanen, J.: Fast adaptive arithmetic code for large alphabet sources with asymmetrical distributions. IEEE Commun. Lett. **7**(1), 33–35 (2003)
39. Ryabko, B.Ya., Stognienko, V.S., Shokin, Yu.I.: A new test for randomness and its application to some cryptographic problems. J. Stat. Plan. Inference **123**, 365–376 (2004)
40. Shimoyama, T., Takeuchi, K., Hayakawa, Ju.: Correlation attack to the block cipher RC5 and the simplified variants of RC6. In: Proceedings AES3, New York (AES3 Paper Submissions). http://csrc.nist.gov/encryption/aes/round2/conf3/aes3papers.html (2001)
41. Soto, J., Bassham, L.: Randomness testing of the Advanced Encryption Standard finalist candidates. In: Proceedings AES3, New York. http://csrc.nist.gov/encryption/aes/round2/conf3/papers/30-jsoto.pdf (2001)

Chapter 3
SCOT-Modeling and Nonparametric Testing of Stationary Strings

3.1 Introduction

Everyone has the experience of using universal compressors (UC) such as Zip for storing and transmitting files. Our group studied their application for **statistical analysis of stationary time series.**[1]

Let $\mathbf{B} = \{0, 1\}$, $x^N = (x_1, \ldots, x_N) \in \mathbf{B}^N$ be a binary (**training**) string having a stationary ergodic distribution (SED) P_0.

C. Shannon [72] created a comprehensive theory of information transmission of strings modeled as stationary stochastic sequences. In particular, **given** a distribution $P(a)$ on a finite alphabet A such that $|\log P(a)| = l(a)$ are integers, let us consider the **mean length (complexity)** of the Shannon-Fano encoding elements $a \in A$ of an Independent Identically Distributed (IID) source $\{a_i \to l(a_i), i = 1, 2, \ldots, a_i \in A\}$. Then the sequence of binary $l(a_i)$-tuples asymptotically attains Shannon's entropy lower bound for the complexity of the compression. Kolmogorov [34] sketched a Turing-machine-based **complexity theory of an individual string without any randomness assumptions** and outlined an idea of the first **Universal Compressor (UC)**, such that for large strings *belonging to an IID statistical ensemble*, their complexity approximates their entropy. This groundbreaking idea was used to create the first UC which adapts to an **unknown** stationary ergodic distribution (SED) of strings and whose complexity asymptotically attains the Shannon entropy lower bound (followed by Lempel-Ziv UC LZ-77, LZ-78 and [67] among others). Let $\mathbf{P}$ be the class of all the SED sources approximated by m-MC's and $\log := \log_2$. Compressor family $\mathbf{L} = \{\lambda_n : \mathbf{B}^n \to \cup_{i=1}^{\infty} \mathbf{B}^i$ is

[1]Several students listed in the Acknowledgement section made significant contributions to what will be described in this part.

B. Ryabko et al., *Compression-Based Methods of Statistical Analysis and Prediction of Time Series*, DOI 10.1007/978-3-319-32253-7_3

(weakly) **universal** if for any $P \in \mathbf{P}$ and any $\epsilon > 0$, the following condition holds:

$$\lim_{n\to\infty} P(x \in \mathbf{B}^n : |\lambda_n(x)| + \log P(x) \le n\epsilon) = 1, \tag{3.1}$$

where $|L_n(x)|$ is the length of $L_n(x)$ and $|\lambda_n(x)| + \log P(x)$ is called *individual redundancy*. Thus, for a string generated by an SED, the **UC-compression length asymptotically is its negative log-likelihood** which can be used in **nonparametric** statistical inference if the *likelihood cannot be evaluated analytically*. Fitingof [20] modeled stationary strings as m-Markov Chains (m-MC, m-grams) and estimated their numerous parameters. Then those parameters are used for compression. Instead of full m-MC, Rissanen [63] uses their less parametrized class VLMC where every state is independent of the states which are more remote than the *contexts* of a certain length depending on the preceding m-gram. We can statistically estimate the Stochastic Context Tree (SCOT) of the training string and then use it for the VLMC-compression.

An arbitrary UC (see [78]) maps the source strings $x^N \in \mathbf{B}^N$ into compressed binary strings x_c^N of length $|x_c^N|$. The probability distribution on the set of all compressed sequences $|x_c^N|$ is induced by the original distribution P over $\mathbf{B}^N$. However, because a UC provides a 1-1 map, the probability of x_c^N in that induced probability distribution coincides with $P(x^N)$. Then by (3.1), $|x_c^N|$ generates the *approximate log-likelihood* of source x^N which we will use as the main inference tool about P.

None of the LZ-compressors use any statistics of strings at all. Instead, LZ-78 consecutively constructs the tree of the binary patterns unseen before starting with the first digit of the string. Ziv and Wyner proved that LZ-78 and LZ-77 are UC entailing:

$$\lim_{n\to\infty} P(|\lambda_n(x)|/|x| \to h) = 1 \tag{3.2}$$

for $P \in \mathbf{P}$, where h is the binary entropy rate (per symbol). As was shown in [73], where SED strings were first singled out as popular models of natural language, h is an asymptotic lower bound for compressing an SED source. Conversely, an elementary argument shows that (3.2) implies (3.1). By the 1990s, various versions of LZ-77 or LZ-78 became everyday tools in computer practice.

The **Minimum Description Length principle (MDL)** [64] and goodness of fit and homogeneity testing [78] initiated UC-based applications to statistical inference about SED sources, including several recent papers of B. Ryabko with coauthors. Of special interest to us is the **homogeneity test** in [68].

Instead of compression, we use SCOT in Sect. 3.5 for generating the likelihood function of strings and apply it for statistical inference about financial and seismological time series, as well as in similar type sequences used in literary research analysis. A substantial part of Sect. 3.3 is devoted to the innovative modeling of SCOT-governed time series. We start with theoretical results on approximate likelihood generated by UC and complete Chap. 3 with statistical applications of UC-based inference to authorship attribution of literary texts.

The structure of Chap. 3 is as follows:

- Section 3.2: Theory of UC-Based Discrimination
- Section 3.3: The SCOT Theory and Modeling
- Section 3.4: Limit Theorems for Additive Functions of SCOT Trajectories
- Section 3.5: SCOT-Based Homogeneity Testing
- Section 3.6: UC-Based Authorship Attribution

3.2 Theory of UC-Based Discrimination

3.2.1 Homogeneity Testing with UC

Let $|A|$ and $|A_c|$ be the lengths of respectively **binary** string A and its compression A_c.

The *concatenated* string $S = (A, B)$ is the string starting with A and proceeding to text B without stop.

3.2.1.1 Ryabko-Astola-Test and U-Statistics

If the entropy rates of strings A and B are significantly different, then it is elementary to prove their inhomogeneity. In the opposite case the homogeneity test statistic ρ from [68] can help. It is

$$\rho = h_n^*(S) - |A_c| - |Q_c|, \tag{3.3}$$

where the empirical Shannon entropy h_n^* of the concatenated sample S (based on m-MC approximation) is defined in their formula (6). The local context-free structure (microstyle) of long (several Kbytes) literary texts (LT) can be modeled sufficiently accurately only by binary m-MC with m not less than several dozen. Thus its evaluation for LT is extremely intensive computationally and unstable for texts of moderate size requiring regularization of small or null estimates for transition probabilities. Therefore, the appropriateness of ρ and equally computationally intensive m-MC-likelihood-based [66] is questionable.

For short LTs the accuracy of SED model may be insufficient, while for very large LTs such as a novel affected by long literary form relations ('architecture' features such as 'repeat' variations), the microstyle describes only a local part of the author's style as emphasized in [13]. Our follow-up analysis in Sects. 3.5 and 3.6 shows that parts of certain words contribute to discrimination between styles with SCOT and CCC (conditional compression complexity). We expect a more interesting SCOT-discrimination between styles using an alphabet comprised of words rather than letters. This work is now in progress. Some authors (e.g. M. Tarlinskaya) deal with annotated texts, and some researches try to include some elements of grammar in their analysis. Such methods are beyond the scope of our approach.

CCC-Test

Suppose we are given a training binary SED string x^N of dimension N with a joint distribution P_0. Let us consider an SED binary query y^M of dimension M whose random components are jointly distributed as P_1, and compare P_1 with probability distribution P_0 over a space of dimension M to test whether the homogeneity hypothesis $P_0 = P_1$ contradicts the data or not.

Let us extract from y^M several **slices** $y_i, i = 1, \ldots, S$, of identical length n in a small distance between consecutive slices to provide approximate independence of slices (in applications of Sect. 3.6 slices follow at no distance at all because of insufficient data length and practically negligible dependence between slices).

Let us consider compressed concatenated strings $C_i = (x^N, y_i)_c$, i.e., make concatenation (x^N, y_i) and apply our compression method to it. Define $CCC_i = |(C_i)_c| - |x_c^N|$.

Our CCC-statistic $\overline{CCC}$ is the average of all CCC_i. We can find $\overline{CCC}$ in the study of very small error probabilities when finding few inhomogeneous inputs among a vast number of homogeneous ones [44]. Homogeneity of *two* texts is tested by the test statistic $\mathcal{T} = \overline{CCC}/s$, where s is the empirical standard deviation of $CCC_i, i = 1, \ldots, S$. An extensive experimentation with real and simulated data in Sect. 3.6 has shown that $\mathcal{T}$ well discriminates between homogeneity and its absence in spite of the lack of knowledge about P_0 and P_1. Here we prove the consistency of test $\mathcal{T}$, CCC-tail optimality and give a sketch of proof of $\mathcal{T}$—asymptotic normality by showing that test $\mathcal{T}$ approximates the Likelihood Ratio Test (LRT) in full generality under certain natural assumptions about the sizes of the training string and the query slices.

The validity of our Quasi-classical Approximation assumption in Sect. 3.2.3 depends both on the compressor that we used and on the binary SED source distribution.

All results are new. All previous attempts [1, 75] to prove the asymptotic normality under the null hypothesis were extremely long and technical and applied to only *one* compressor and to an IID source of dubious importance in applications.

The main advantages of CCC-test are:

(a) its applicability for arbitrary UC and long memory sources, where the likelihood is hard to evaluate and
(b) its computational simplicity (as compared to statistic ρ and the one from [78]). This simplicity enables online simultaneous processing of multichannel change-point detection, see [44].

3.2.2 Preliminaries

Conditions on regular stationary ergodic distributions (SED) of strings enabling approximation by m-Markov Chains are as in [78].

A.N. Kolmogorov spent decades on attempts to correct the von Mises erroneous criterion of string randomness. As a result, he gave a sketch of a version of Theorem 3.2.1 for IID sources, connecting for the first time the notions of *complexity and randomness* [34]. He also proposed there a compressor adapting to an unknown IID distribution in such a way that (3.2) holds. Influenced by these ideas, the first practically implementable UCs LZ-77/78 were invented within 12 years after that and became everyday tools of computer work after ten more years.

Introduce $L_j^n = -\log P(y^n), j = 0, 1$, under the distribution of the training/query strings.

The entropy $h_j^n = -n^{-1} \sum P(y^n) \log P(y^n), j = 0, 1$, under the distribution of the training/query strings.

Warning Without mentioning it, we actually consider the *conditional probabilities* P_1 *and expectations* E_1 *of functions of the query slice with regard to the training text.*

Kraft Inequality Lengths of every uniquely decodable compressor satisfy the following inequality: $\sum_{\mathbf{B}^n} 2^{-|x_c^n|)} \leq 1$.

Introduce diversion (cross-entropy) $D(P_1||P_0) = n^{-1}E_1 \log(P_1/P_0)$ and consider goodness of fit tests of P_0 vs. P_1.

Theorem 3.2.1 *for the particular case of IID sources is a cornerstone of Kolmogorov's almost life-long attempts to correct von Mises's erroneous randomness criterion.*

'Stein Lemma' for SED [78] *If* $D(P_1||P_0) \geq \lambda > 0$ *and any* $0 < \varepsilon < 1$, *then the error probabilities of LRT satisfy simultaneously*

$$P_0(L_0 - L_1 > n\lambda) \leq 2^{-n\lambda}$$

and

$$\lim P_1(L_0 - L_1 > n\lambda)) \geq 1 - \varepsilon > 0.$$

No other test has both error probabilities less in order of magnitude.

Theorem 3.2.2 ([78]) *Consider test statistic* $T = L_0^n - |x_c^n| - n\lambda$. *The nonparametric goodness of fit test* $T > 0$ *has the same asymptotics of the error probabilities as in the Stein lemma.*

3.2.3 Main Results

We denote $(\lambda_0^N(y)$ the y_c compressed with the encoding generated by the training string x^N.

'Quasi-Classical Approximation' Assumption (QAA) *The sizes of the training string N and query slices n grow in such a way that the joint distribution of* $CCC_i, i = 1, \ldots, S$, *converges in probability to* $P_1(\lambda_0^N(y))$ *(see 3.1) as* $N \to \infty$, *and* $n \to \infty$ *is sufficiently smaller than N.*

The intuitive meaning of this assumption is: given a very long training set, continuing it with a comparatively small query slice with alternative distribution P_1 does not affect significantly the *encoding rule*. The typical theoretical relation between lengths as discussed in [56] is $n \leq const \log N \to \infty$.

In practice, an appropriate slice size is determined by empirical optimization.

Define entropy rates $h_i = \lim h_i^n, h_i^n = E_j L_i^n, i = 0, 1$.

Under QAA and

$$h_1 \geq h_0, \tag{3.4}$$

the following three statements are true.

Theorem 3.2.3 (Consistency) *The mean CCC becomes strictly minimal as* $n \to \infty$ *if* $P_0 = P_1$.

Proof $E_1(CCC) - E_0(CCC) = \sum(P_0(x) - P_1(x)) \log P_0(x) = -h_0^n + \sum P_1(x) \log P_1(x)(P_0/P_1) = h_1^n - h_0^n + D(P_1||P_0)$. Proof now follows from (3.4) and positivity of divergence unless $P_0 = P_1$. If $h_1 < h_0$ (say, G. Chapman versus C. Marlowe), then the inhomogeneity can be proved by estimating both conditional and unconditional complexities of texts.

Generate an artificial n-sequence z^n independent of y^n and P_0-distributed, and denote by CCC_0 its CCC.

In Sect. 3.6 such a test was obtained by training the slice on the remaining portion of the same text. Also assume that the distances between slices are such that the joint distribution of S slices of size n converges to their product distribution in probability.

Theorem 3.2.4 *Suppose* P_1, P_0 *are SED,* $D(P_1||P_0) > \lambda > 0$ *and we reject homogeneity if the 'conditional version of the Likelihood Ratio' test* $\mathcal{T} = \overline{CCC} - \overline{CCC_0} > n\lambda$. *Then the same error probability asymptotics as for LRT is valid for this test.*

Proof Sketch Under independent slices, their probabilities multiply. To transparently outline our ideas (with some abuse of notation) replace the condition under summation sign to a similar one for the whole query string: instead of $P_0(T' > 0) = \sum_{y,z:\overline{CCC}-\overline{CCC_0}>\lambda} P_0(y)P_0(z)$, we write the condition under the summation as $CCC - CCC_0 > n\lambda$, which is approximated in probability under P_0 by inequality $L_n(y) - |z| > n\lambda$.

Thus, $P_0(T' > 0) \leq \sum_z \sum_{P_0 \leq 2^{-n\lambda - |z|}} P_0(y)P_0(z) \leq 2^{-n\lambda} \sum_z 2^{-|z|} = 2^{-n\lambda}$ by the Kraft inequality. We refer to [78] for an accurate completion of our proof in a similar situation.

Informally again, $\lim P_1(T' > 0) = \lim P_1(n^{-1}(|y| - |z|) > \lambda = D(P_1||P_0) + \varepsilon, \varepsilon > 0$. $|y|/n$ is in probability P_1 around $-\log P_0(y) = E_1(-\log(P_0(y))) + r$, $|z|/n$ is in probability P_0 around $-\log P_0(z) = h_0^n + r'$. As in the Consistency proof, all the principal deterministic terms drop out, and we are left with the condition $r < \varepsilon + r'$. Its probability converges to 1 since both r, r' shrink to zero in the product (y, z)-probability as $n \to \infty$.

Let us study the central range of CCC-distribution and now assume in addition to QAA that P_1 *is contiguous w.r.t.* P_0, i.e. the sets of P_0 measure 0 have also P_1 measure 0. This assumption for m-MC means that transitions which are impossible under P_0 are equally impossible under P_1. This is natural for LT in the same language. We assume also that P_0-distribution of L^n is asymptotically Normal (AN). The normal plots to every application in Sect. 3.6 are amazingly strong empirical confirmations for the AN distribution of our CCC-statistic.

Theorem 3.2.5 *AN holds also for* $CCC_i, i = 1, \ldots, S$, *under* P_1 *and all assumptions made (Le Cam lemma [30]). Statistic* $\mathcal{T}$ *has asymptotically central/non-central Student distribution respectively under* P_0 *and* P_1 *with* $S - 1$ *degrees of freedom.*

3.2.4 Sketch of AN Justification Under the Null-Hypothesis

In IID cases, AN is proved in [1, 75] on several dozen pages that make for hard reading. Our approach for general UC and SED outlined below is shorter.

J. Ziv's claim (personal communication)—Since y^n is almost incompressible, the compressed file right hand tail with infinite/long memory is (respectively, converges to in divergence) IID(1/2). Thus, the length of compressed texts is an asymptotically sufficient statistic containing all information about the training SED, while the compressed text itself is pure noise!

We have UC: $y^n \to Z_n := (m(n), z^m)$, The joint distribution of the random vector Z_n is $\mathcal{L}_m$. Given the training text, Z_n is a deterministic invertible function of y^n. Thus, entropy $H(\mathcal{L}_m) = h^n$.

According to the Ziv's claim, $h^n/m(n) \to 1$ in probability. 'In probability' will further mean in probability of the joint distribution of x^N, y^n, where for some $g(\cdot) :\to \mathbf{N}, n < g(N)$.

Assumption A: **'Second Thermodynamics Law'** UC is such that $h_N^n/m_N(n)$ growth is monotone in probability. The limiting joint distribution $\mathcal{L}_{m(x^n)}$ maximizes entropy $H(x^n))$ over the set $\{Var(h^n) \geq const\}$ in probability.

A Heuristic Commentary About Validity of A Define P_N, P_∞-conditional distributions $y_c^n = z^m$ under large N and $N = \infty$, $D^* = \lim D/n, h^* = \lim h/n$.

UC LZ-78 constructs the pattern tree rooted at $-\infty$, its right hand end consists of sub-trees of the length $N+n$—training text of length N and slices of length n. The conditional probability in this half-infinite random tree is $\lim \frac{\log P_\infty(P_N \in G)}{n}$.

If we prove the applicability of Sanov's Large Deviations theorem [16] for $G = G_N = \{Var(h^n) \geq const\}$ or a somewhat different G with the same meaning, then Sanov's theorem gives:

$$\lim \frac{\log P_\infty(P_N \in G)}{n} = -\min_{P_N \in G} D^*(P_N || P_\infty) = \max_{P_N \in G} h^*(P_N) - 1/2,$$

which shows that the typical N-approximations to P_∞ must belong to $\{P_N : h(P_N) = \max_{P_N \in G} h(P_N)\}$.

Corollary *$h^n/m(n)$ is a nondecreasing submartingale under A bounded by 1; it is AN in probability for large n after removing a fitted small parabolic drift; all random IID parameters of the Bernoulli variables z_i are AN due to the well-known entropy maximization of a shrinking distribution on* $[0, 1]$.

Now, $m(n)$ is the first exit time when the AN sequence $s^i = \sum_1^i h(z_j)$ hits high level h^n. Its AN is well known, see [29].

3.2.5 Preceding Work

Kolmogorov's Theorem 3.2.1 on complexity and randomness relation is the cornerstone in our constructions. Being partly a philosopher, on the eve of his final grave illness, he made a sketch of complementing this with the Principle of Kolmogorov Complexity (KC-AT) based on Turing's pseudo-algorithmic theory. This was done in parallel to Solomonov and Cheitin, who were far from Mathematics. The KC-AT inspired D. Khmelev to introduce a core of our CCC-method *outside of the statistical paradigm*. For an SED string x^N and a fixed UC, $|x_c^N|$ is an approximation to the conditional KC. However, if x^N is a genome segment, $|x_c^N|$ is several times more than $|x^N|$ for UC zip and is in no way an approximation to abstract conditional KC. I am unaware of a nontrivial sufficiently rich class of cases in addition to statistical models where *incomputable* KC can be approximated from below and above by *computable functions*, at least theoretically. Thus, replacement of KC with a quantity evaluated with UC as in [14] needs justification, to say the least. Being far from elementary, statistical UC theory *must* be applied for strings approximated by a statistical model such as LT.

A survey of related early developments (before [41] first introduced a working version of CCC) is in [14]. All of them follow Khmelev, adding artificial transformations as in [14] irrelevant in statistical context and only worsening accuracy of analysis; their replacement of KC with quantity evaluated with UC is not justified. Thus the validity of their applications is doubtful. Their classifier poorly discriminates between the LTs in the same language which are significantly

CCC-discriminated (see [43]) and their output depends on texts entropies not mentioned at all in [14]. Their claim that L. Tolstoy stands alone among Russian writers is caused most likely by inadequate text preparation for analysis: they did not remove large portions of French in Tolstoy having a different (from Russian) entropy rate.

3.3 SCOT-Modeling

Modeling random processes as full m-Markov Chains (m-MC) can be inadequate for small m, and over-parametrized for large m. For example, if the cardinality of the base state space is $4, m = 10$, then the number of parameters is around 3.15 million. The popular Box–Jenkins ARIMA and Engel's GARCH in quality control and finance respectively are not adequate in applications to linguistics, genomics and proteomics, security, etc., where comparatively long *non-isotropic contexts* are relevant that would require huge memory size for the full m-MC.

The popularity of another compression tool—*sparse* Variable Memory Length MC (VLMC), based on SCOT fitting, has been increasing rapidly since a fitting algorithm was proved to be *consistent for stationary mixing sequences* in [63] and used for compression. Stochastic COntext Tree (abbreviated as SCOT) is an m-Markov Chain (m-MC) with every state independent from all states preceding the **context**. The **length** of context is determined by the preceding symbols to this state. SCOT has also appeared in other fields under the somewhat confusing names VLMC, VOMC, PST, CTW, among others, *for compression applications*. Apparently, the first SCOT *Statistical Likelihood* comparison application [4] and rather complicated statistical tests on stochastic suffix trees in [2, 12] to *highly non-stationary Bioinformatics data* are inadequate.

Sparse SCOT over some alphabet A is a very special case of m-MC, where m is the maximal length of **contexts**.[2]

With a given string $x_{-m}, \ldots, x_{-1}, x_0$, the context

$$C(x_{0-}) = x_{-1}, \ldots, x_{-k}, k \leq m := x_{-1}^{-k} \quad (3.5)$$

(to a current state x_0 given preceding m-gram) is the end segment of the preceding m-gram of **minimal length** such that the conditional probability

$$P(x_0|x_{-1}^{-r}) \equiv P(x_0|x_{-1}^{-k}), \forall r > k; k = |C(x_{0-})| \quad (3.6)$$

is called the length of context $C(x_{0-})$.

[2]Galves and Loecherbach [24] and some other publications of dubious practical importance admit $m = \infty$.

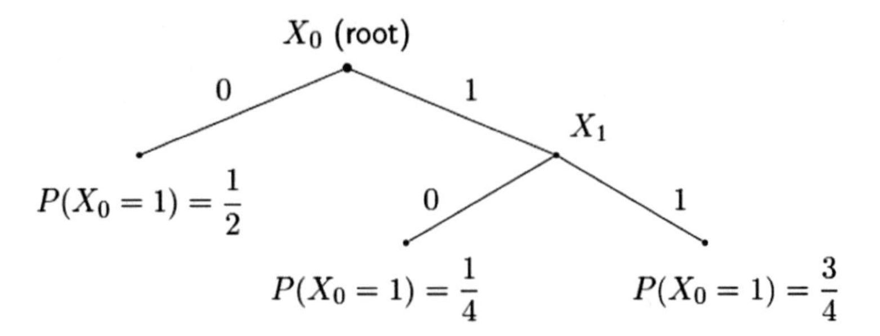

Fig. 3.1 The simplest stochastic context tree (Model 1)

For large m, SCOT is sparse if the total number of contexts is $O(m^k)$ for some fixed k. Note that symbols of contexts are written in the natural order starting from the oldest one. The **same number** $|A|$ **of edges** goes from any node of the context tree other than the context end (leaf) which is valid for SCOTs obtained by the training algorithm and crucial for our lemmas in the next section.

The simplest SCOT model with alphabet $\{0, 1\}$, contexts $\{\mathbf{0}\}, \{\mathbf{01}\}, \{\mathbf{11}\}$ and corresponding transition probabilities $P(x_0 = 1)$ given preceding contexts are respectively $1/2, 1/4$ and $3/4$ as displayed in Fig. 3.1. Further, we find SCOT stationary distribution in this and several more models to prepare for more advanced topics: invariance, asymptotic normality and exponential tails of additive functions which we prove in Sect. 3.3. These are to validate our statistical analysis in [47] and our Sect. 3.5, which showed an advantage of SCOT approximation over the popular GARCH in distinguishing between quiet and volatile regions of certain financial time series (FTS). The reason for this advantage seems to be an anisotropic memory required to predict FTS. We finish Sect. 3.3 by choosing parameters of our last SCOT model to match the historical Apple FTS.

3.3.1 m-MC Reduction to MC on A^m

An m-MC $\{X_n\}$ with a finite state space (alphabet) $A = \{a_1, \ldots, a_d\}$ can be regarded as 1-MC

$$\{Y_n = (X_n, X_{n+1}, \ldots, X_{n+m-1})\}$$

with alphabet as the space of all *m-grams* A^m : $P(Y_{n+1}|Y_n) = P(X_{n+m}|Y_n)$, if $X_{n+1}, \ldots, X_{n+m-1}$ coincide on both sides, and 0 otherwise.

This induced MC on A^m is not necessarily ergodic for ergodic m-MC. A simple counterexample follows.

3.3.2 Counterexample

Consider a binary 2-MC with alphabet $\{0, 1\}$ and transition probability 1/2 from $\{0, 1\}$ or $\{1, 1\}$ to $\{0\}$ and $\{1\}$, transition probability 1 from $\{0, 0\}$ and $\{1, 0\}$ to $\{1\}$. It is ergodic, but the induced MC on 2-grams has transient state $\{0, 0\}$.

3.3.3 1-MC Model Induced by SCOT

Let T be a context tree; C(T):= T^* denotes the set of contexts in T, T_i denotes the subtree T_i of T whose root is a_i, thus $T_i^* := \{u|\ \overline{ua_i} \in T^*\}$. For all pairs $a_i \in A, a_j \in A$, $T_{j,i} := T_i(T_j)$, thus $T_{j,i}^* := \{u|\overline{ua_ia_j} \in T^*\}$ is the concatenation of u, a_j in the natural ordering.

To implement the idea of Sect. 3.3.1 for SCOT models, it is sufficient to verify that after moving from x_i to $x_{i+1}, i = 0, \ldots,$ the context for the latter will **not include symbols older than in C(**x_{i-}**).**

This is formalized by the following

Definition A complete context tree T is called "TailClosed" if $\forall\ c \in T^*, i \in \{1, \ldots, d\}$, $\exists\ u \in T^*$, s.t. $\overline{ca_i} = \overline{wu}$, where w is a string.

Theorem 3.3.1 (Necessary and Sufficient Condition, T. Zhang) *Let T be a complete context tree, then the following statements are equivalent.*

(a) *T is TailClosed.*
(b) $\forall\ 1 \leq i, j \leq d,\ T_{j,i} \subseteq T_i$.
(c) $\forall\ 1 \leq i, j \leq d,\ c \in T_{j,i}^*,\ \exists\ c' \in T_i^*,$ *s.t.* $c' = \overline{uc}$, *where u is a string.*
(d) $\forall\ 1 \leq i, j \leq d,\ c' \in T_i^*,\ \exists\ c \in T_{j,i}^*,$ *s.t.* $c' = \overline{uc}$, *where u is a string.*

T. Zhang also constructed a TailClosed envelope of arbitrary SCOT model.

Note that every SCOT T must contain at least one complete set of 'reversed siblings', that is, they are different only in the first symbol (from the bottom). Denote by T' the tree obtained by replacing a set of 'reversed siblings' with one node t'.

Let $M(T)$ denote the maximal difference between context sizes in T^*.

Theorem 3.3.2 (Sufficient Condition) *If $M(T) \leq 1$, then the SCOT is TailClosed.*

The following auxiliary statements hold:

Lemma 3.3.1 $M(T') \leq 1 \leftrightarrows M(T) \leq 1$. *Proof is straightforward.*

Lemma 3.3.2 *1. If $C(T)$ is TailClosed, then $C(T')$ is also TailClosed.*

Proof is given only for binary alphabet to simplify the notation. Let $\overline{0, X_1, X_2, \ldots, X_m}$ and $\overline{1, X_1, X_2, \ldots, X_m}$ be 'reversed siblings'. Fix arbitrary string z such that length $|z| \geq \min_{t \in C(T)} |t|$.

There exists $t \in C(T)$ such that $z = at$. If $t \in \overline{0, X_1, X_2, \ldots, X_m} \cup \overline{1, X_1, X_2, \ldots, X_m}$, let $t' = \overline{X_1, X_2, \ldots, X_m}$.

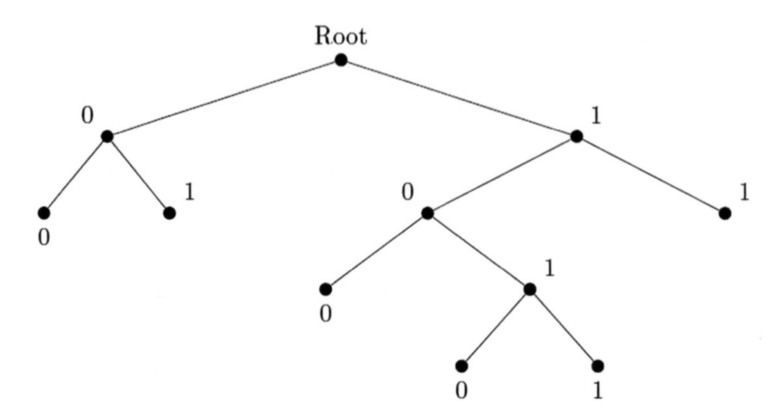

Fig. 3.2 Counterexample

We have $z = \overline{a't'}$, and $t' \in C(T')$. If t is in the complement to $\overline{0, X_1, X_2, \ldots, X_m} \cup \overline{1, X_1, X_2, \ldots, X_m}$, then $t \in C(T')$ itself.

Lemma 3.3.3 *Every binary context tree T such that $M(T) = 0$ is TailClosed. Proof is straightforward.*

Proof of the Theorem Let the maximal context size be m. If $M(T) = 0$, our statement follows from Lemma 3.3.3. If $M(T) = 1$, then there must be a sequence $\{T_i\}$ of binary context trees in T such that $M(T_1) = 0$ (say, we start with the full m-MC), and the maximal context size of T_1 is m; and T_{i+1} is obtained from T_i by cutting off the longest sibling leaves. By Lemma 3.3.3, T_1 is TailClosed. By Lemma 3.3.2, $T_1, T_2, \ldots, T$ are all TailClosed.

Counterexample (See Fig. 3.2) Consider C(T)=00,10,001,0101,1101,11.

If x_0=1 and $C(x_{0-})$=((10)), then (10)1=101, which is not a context.

Generally, the induced MC are found by direct computations.

3.3.4 Stationary Distribution for Model 1

Similarly to the full m-MC case, we adopt the following **definition**:

The transition probability from C(x_{i-}) to C(x_{i+1-}) is *by definition* the transition probability from C(x_{i-}) to x_{i+1} if SCOT is TailClosed.

In our first example, the above definition leads to the following transition matrix **P** between contexts:

	11	**0**	**01**
11	0.75	0.25	0.0
0	0.00	0.50	0.5
01	0.25	0.75	0.0

This MC is ergodic. We evaluate its stationary distribution—row-vector $\mathbf{Q} = (q(\mathbf{11}), q(\mathbf{0}), q(\mathbf{01}))$ of contexts which is the unique solution to the equation $\mathbf{Q} = \mathbf{QP}$. This solution for this elementary example can be found by hand. In more advanced cases, the following iterative procedure exists: use a not bad preliminary guess $\mathbf{Q}^0$ and multiply $\mathbf{Q}^0$ by $\mathbf{P}^n$ for large n. Due to the well-known Perron-Frobenius theorem about the exponential convergence of these products to $\mathbf{Q}$ as $n \to \infty$, we approximate $\mathbf{Q}$ with arbitrary precision.

$\mathbf{Q}$ turns out to be $(1/4, 1/2, 1/4)$.

Although we showed a counterexample to the availability of SCOT reduction to MC on the space of contexts in the preceding subsection, this reduction seems to be generally valid, as shown by two following examples neither of which satisfies the sufficient condition above. Thus, reduction possibility to MC and its ergodicity are either validated or assumed, when appropriate.

3.3.5 'Comb' Model D_m

Binary Context Tree D_m repeats m times the splitting in the right-hand side of Fig. 3.1. It has contexts $(0), (01), (011), \ldots, (01^{m-1}), (1^m)$ and admits reduction to 1-MC for every m. Let us assign all root probabilities of $\{0\}$ and $\{1\}$ given every context as 1/2.

Then this SCOT is ergodic, the **stationary distribution** $\{q_i\}$ for this 1-MC is $(1/2, 1/4, \ldots, (1/2)^{m-1}, (1/2)^m, (1/2)^m)$, and the entropy rate

$$ER = -\sum_{i \in A} \sum_{j \in A} q_i p_{ij} \log p_{ij}$$

is log 2 for all m.

3.3.6 *Models 2*

We start with over-simplified models to warm up and then move towards more realistic models.

Model 2 (i)

See Fig. 3.3.

This model leads to an undesirable periodicity of the MC over contexts which we later fix by introducing a tiny positive probability of staying at every state.

$X_n = X_{n-1} + 1$ with probability 0.9, if $X_{n-1} = X_{n-2} + 1$ and $X_{n-1} \neq \pm l$;

X_n is defined anti-symmetrically to the above, if $X_{n-1} = X_{n-2} - 1$ and $X_{n-1} \neq \pm l$.

Finally, the process reflects one step from the boundary next time after hitting it. Thus, the transition probability matrix is:

	−2.down	−1.up	−1.down	0.up	0.down	1.up	1.down	2.up
−2.down	0	1	0	0	0	0	0	0
−1.up	0.1	0	0	0.9	0	0	0	0
−1.down	0.9	0	0	0.1	0	0	0	0
0.up	0	0	0.1	0	0	0.9	0	0
0.down	0	0	0.9	0	0	0.1	0	0
1.up	0	0	0	0	0.1	0	0	0.9
1.down	0	0	0	0	0.9	0	0	0.1
2.up	0	0	0	0	0	0	1	0

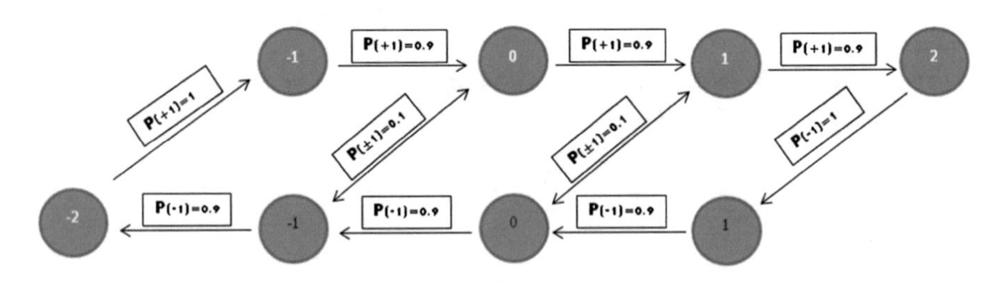

Fig. 3.3 Model 2 (i) is displayed

The powers of transition probability matrix converge after even number of steps to:

	−2.down	−1	0	1	2.up
−2.down	0.25	0	0.5	0	0.25
−1	0	0.5	0	0.5	0
0	0.25	0	0.5	0	0.25
1	0	0.5	0	0.5	0
2.up	0.25	0	0.5	0	0.25

The powers of transition probability matrix converge after odd number of steps to:

	−2.down	−1	0	1	2.up
−2.down	0	0.5	0	0.5	0
−1	0.25	0	0.5	0	0.25
0	0	0.5	0	0.5	0
1	0.25	0	0.5	0	0.25
2.up	0	0.5	0	0.5	0

Model 2 (ii)

See Fig. 3.4.

In Model 2 (ii), we modify the previous one by introducing a probability of staying at the same state to avoid periodicity:

$$X_n = \begin{cases} X_{n-1} + 1, \text{with prob } 0.8, \\ X_{n-1} - 1, \text{with prob } 0.1, \ \text{if } X_{n-1} = X_{n-2} + 1 \text{ and } X_{n-1} \neq \pm l, \\ X_n - 1, \text{with prob } 0.1. \end{cases}$$

$$X_n = \begin{cases} X_{n-1} + 1, \text{with prob } 0.1, \\ X_{n-1} - 1, \text{with prob } 0.8, \ \text{if } X_{n-1} = X_{n-2} + 1 \text{ and } X_{n-1} \neq \pm l, \\ X_n - 1, \text{with prob } 0.1. \end{cases}$$

$$X_n = \begin{cases} X_{n-1} + 1, \text{if } X_{n-1} = -l\,, \\ X_{n-1} - 1, \text{if } X_{n-1} = l\,. \end{cases}$$

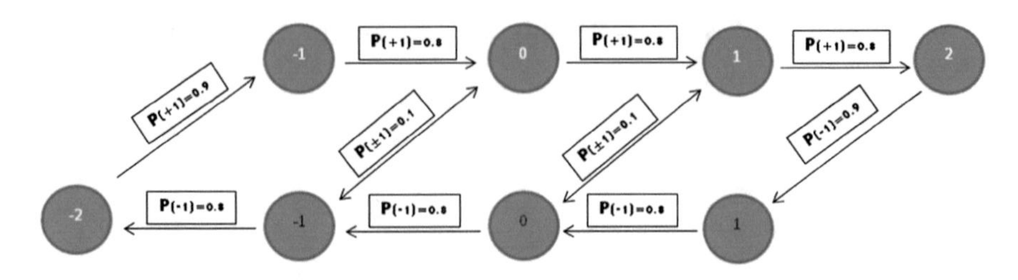

Fig. 3.4 Illustrated is model 2 (ii)

The transition probability matrix is:

	−2.down	−1.up	−1.down	0.up	0.down	1.up	1.down	2.up
−2.down	0.1	0.9	0	0	0	0	0	0
−1.up	0.1	0.1	0	0.8	0	0	0	0
−1.down	0.8	0	0.1	0.1	0	0	0	0
0.up	0	0	0.1	0.1	0	0.8	0	0
0.down	0	0	0.8	0	0.1	0.1	0	0
1.up	0	0	0	0	0.1	0.1	0	0.8
1.down	0	0	0	0	0.8	0	0.1	0.1
2.up	0	0	0	0	0	0	0.9	0.1

The powers of the transition probability matrix converge to:

	−2.down	−1	0	1	2.up
−2.down	0.125	0.25	0.25	0.25	0.125
−1	0.125	0.25	0.25	0.25	0.125
0	0.125	0.25	0.25	0.25	0.125
1	0.125	0.25	0.25	0.25	0.125
2.up	0.125	0.25	0.25	0.25	0.125

The stationary distribution is $(0.125, 0.25, 0.25, 0.25, 0.125)$, which is $\left(\frac{1}{2n}, \frac{1}{n}, \ldots, \frac{1}{n}, \frac{1}{2n}\right)$, where $l = 2n$.

Model 3 (i)

When $X_{n-1} \neq \pm l$,

$$X_n = \begin{cases} X_{n-1} + 1, \text{with prob } 0.9, \text{if } X_{n-1} = X_{n-2} + 1 \text{ or } X_{n-1} = X_{n-2} = X_{n-3} + 1, \\ X_{n-1}, \text{with prob } 0.1. \end{cases}$$

$$X_n = \begin{cases} X_{n-1} - 1, \text{with prob } 0.9, \text{if } X_{n-1} = X_{n-2} + 1 \text{ or } X_{n-1} = X_{n-2} = X_{n-3} - 1, \\ X_{n-1}, \text{with prob } 0.1. \end{cases}$$

$$X_n = \begin{cases} X_{n-1} + 1, \text{with prob } 0.05, \\ X_{n-1} - 1, \text{with prob } 0.05, \ \text{if } X_{n-1} = X_{n-2} = X_{n-3} - 1 \\ X_n - 1, \text{with prob } 0.9. \end{cases}$$

When $Xn - 1 = \pm l, X_n = X_{n-1} - \frac{X_{n-1}}{|X_{n-1}|}$.

Model 3 Distribution

Following the model described above and using an initial state ($x_1 = 0, x_2 = 1, x_3 = 2$) of MC with l=15 (31 states in total) we calculated X_{500} simulated 10,000 times. We got the following result:

−15	−14	−13	−12	−11	−10	−9	−8	−7	−6
0.0713	0.0349	0.0316	0.0243	0.0305	0.0288	0.0294	0.0258	0.0296	0.0248

−5	−4	−3	**−2**	−1	0	1	2	3	4
0.0265	0.0279	0.0297	0.0316	0.03	0.0315	0.0273	0.0292	0.0274	0.0253

5	6	7	8	9	10	11	12	13	14	15
0.0305	0.0296	0.0301	0.0302	0.0296	0.0294	0.0273	0.0287	0.0277	0.0389	0.0806

Figure 3.5 is a chart describing the above results.

It is plausible from this chart that the stationary distribution is uniform inside $(-l,)$. To prove this, one can introduce the row $2l + 2$-vector with all inner entries equal to 1 and two boundary entries x. Multiplying it by P from the right, one obtains the proportional vector for appropriate x if this conjecture is correct.

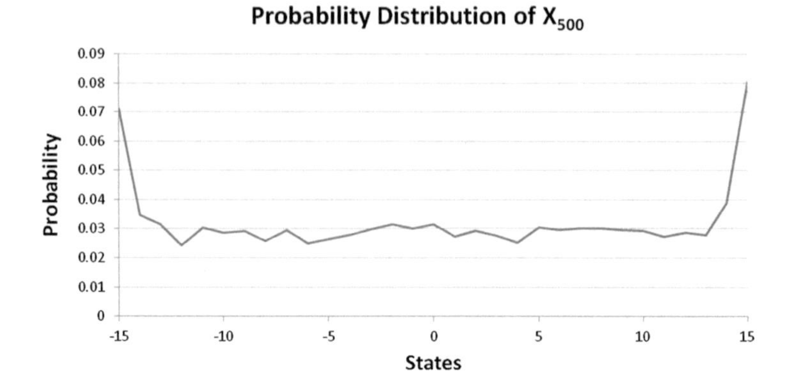

Fig. 3.5 The probability distribution of X_{500} for the different states

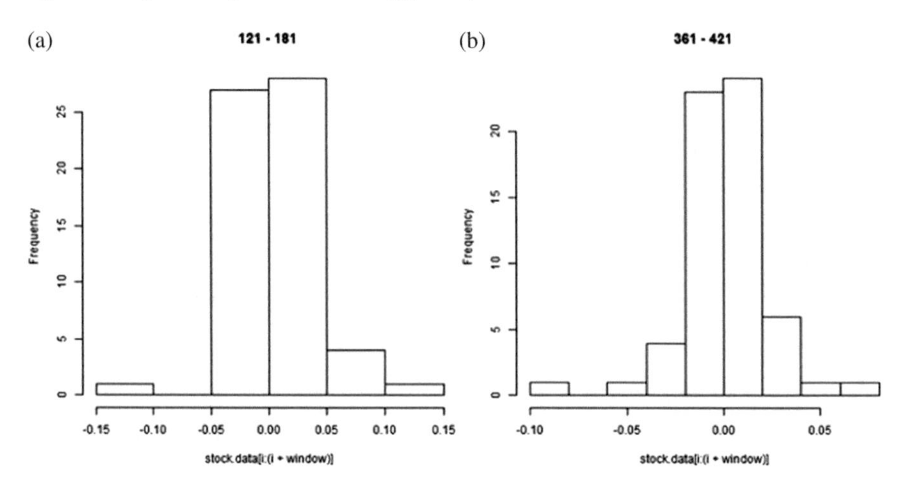

Fig. 3.6 (**a**) The 60-day time-interval 121–181; (**b**) the 60-day time-interval 361–421

Oversimplified Modeling of Financial Data as Model 3

Financial log-returns take arbitrary real values. Nevertheless, it might be worthwhile to model them as Model 3 with parameters matching the data. For doing this, we plot histograms of first and second differences of log-returns with appropriate bins. The Apple data was taken from a two-year period and then the differences and second differences of the log-returns were evaluated of the adjusted closing price—this would be the second derivative on the log-returns of the closing price. Then we split the data into 60-day periods and the "centrality" of each 60-day window was checked via a histogram for frequency of variation. The 60-day window is the title of each histogram shown in Figs. 3.6 and 3.7 in this section.

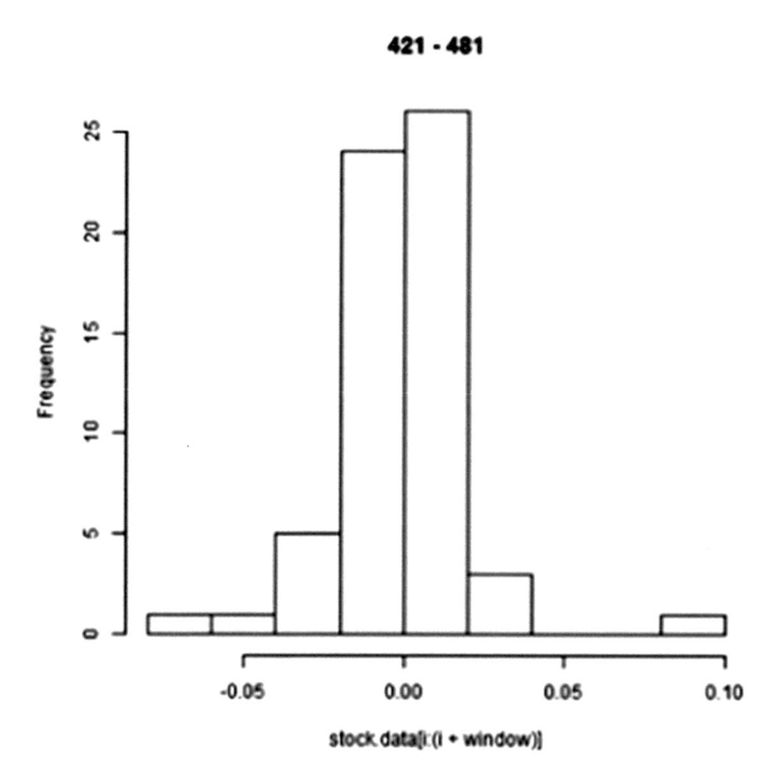

Fig. 3.7 The 60-day time-interval 421–481

3.4 Limit Theorems for Additive Functions of SCOT Trajectories

The asymptotic normality (AN) of additive functions is studied first for ergodic finite alphabet m-MC trajectories. We discuss then what improvements can be obtained for the particular case of sparse SCOT models. Our apparently new simple results for m-MC follow in a straightforward way from those for 1-MC (further called MC). The AN for MC has been known under appropriate conditions since S.N. Bernstein's works around a century ago and strengthened in numerous further publications. For readers' convenience, we refer mostly to the popular comprehensive review [57] available online. We discuss in some detail the Large Deviations (LD) of additive functions only for a particular case of discrimination between two composite hypotheses as applied to Separate Testing of Inputs in search of active inputs of a *sparse system with stationary noise*. The error probability under activeness is fixed in this application, while the null hypothesis error should be as small as possible similarly to [46]. More general MC LD results known for almost half a century are surveyed e.g. in [70], also available online.

3.4.1 SCOT Models Admitting Continuous Time Limit

Ergodic and non-ergodic ladder caricatures of the 'Galileo inertia law with friction' were analyzed in Sect. 3.3.6. Unfortunately, our attempts to construct continuous time limits for them failed. The motion of a Brownian particle hitting heavy objects at random 'Spike'-times is described by another SCOT—'Spike' model which is defined in two steps. The first step is to assign randomly the increments of the random walk to *regular ones* with probability $1 - 2/N$, and (with $2/N$ probability) to spikes. The second step is specifying standard increment distribution for regular increments of ± 1 and a SCOT model with increments of magnitude either 0, or $\pm\sqrt{N}$ to spikes. Here we sketch the functional convergence proof of the Spike model to a martingale—*mixture of the Brownian motion and a symmetric pair of* $\pm$ *Poisson processes.*

The family $X_n^N = \sum_{i=1}^n r_i$ of Spike models is in few words a reflected Random Walk on large interval $[-l, l], l > N^{3/2}$. The regular part of X_n^N has increments ± 1 and reflects one step from the boundary next time after hitting it. Very rare (probability $2/N$) random interruptions by spikes at random Spike time moments n have magnitudes 0 or $\pm\sqrt{N}$ depending on whether $X_n^N = X_{n-1}^N$ (this event can happen only after a very unlikely reflection of a spike from the boundary to the same point) or $X_n^N > X_{n-1}^N$, or the opposite inequality holds. More formally, $X_n^N = \sum_{i=1}^n r_i, r_i$ *in a regular part* is an equally likely sequence of independent identically distributed (IID) $\pm 1, i = 1, \ldots,$ inside $(-l, l)$, while the *irregular part* is a SCOT model specified above.

3.4.2 Continuous Time Limit

Let the increments of time/space be respectively $1/N, 1/\sqrt{N}$ instead of 1. Introduce $w^N(t) = N^{-1/2} X^N_{\lfloor Nt \rfloor}$ (summation is until the integer part $\lfloor Nt \rfloor$ of Nt). We study the weak convergence of $w^N(t)$ as $N \to \infty$.

Inside $(-l, l)$ *conditionally on no spike at time* $k + 1$

$$E(X_{k+1} - X_k) = 0,$$

$$\mathrm{Var}(X_{k+1}^N - X_k^N)) = (1 - 2/N)/N.$$

Let τ_k be the k-th spike time. Obviously, $\tau_k - \tau_{k-1}$ are IID, independent of the σ-algebra spanned by $(x_j, j < k - 2)$ converging to the exponential distribution with mean 2.

Theorem 3.4.1 *In the limit we get a weak convergence of $w^N(t)$ to the Wiener process $w(t)$ in between the compound Poisson spikes processes of equally likely* *magnitudes* ± 1 *independent of* $w(t)$*;*

$$P(\tau_1 > t) = \exp(-t/2),$$

where τ_1 *and* $\{x_t\}, t < \tau_1$*, are independent.*

Proof Sketch Follows in a straightforward way along the lines of the familiar proof of standard random walk weak convergence to the Wiener process, including

- establishing convergence of the Finite Dimensional Distributions (FDD) from the convergence of their multivariate characteristic functions,
- verifying the Kolmogorov Uniform Continuity (KUM) of trajectories criterion *in between spikes* by checking that

$$|w^N(t+h) - w^N(t)|^4 \leq const|h|^2,$$

- applying the Prokhorov theorem to the tight family of distributions of X_n^N trajectories *in between spikes*.

Denote the events: $\{\pm(k+1)\} = \{X_k - X_{k-1} = \pm 1\}$. If a spike happens at time $k+1$, then $X_{k+1} - X_k|\{\pm(k)\} = \pm\sqrt{N}$. As $N \to \infty$, this dependence of the preceding event $X_k - X_{k-1}$ becomes negligible, and the sign of the spike becomes independent of the limiting $w(t)$.

3.4.3 'Thorny' $TH_{a,b}$ SCOT Model

Our next model is similar to the Spike model, except that rare random time moments of spikes of magnitude $\pm aN^b$ with similar dependence of spikes magnitude on the past take place with probability $N^{-2b}, 0 < b < 1/8$. In the same limiting situation of time intervals $1/N$ and steps $1/\sqrt{N}$, the KUM criterion is valid with similar parameters; thus trajectories of the limiting $TH_{a,b}$ model are continuous.

Let the martingale sequence $w^N(t)$ be as above. Then $Er_i = 0, Var[w^N(t)] = Nt[(a^2N^{2b-1})N^{-2b} + (1 - N^{-2b})] \to a^2 + 1$. The equality of summands preceding a spike can be neglected. The covariance of $w^N(t)$ converges to that of $\sqrt{(a^2+1)}w(t)$ in a similar way. Thus this model gives larger volatility without noticeable drift in the limit to continuous t. The weak FDD convergence to that of $\sqrt{a^2+1}w(t)$ is valid since the martingale version of the Lindeberg condition holds, see [38]. Thus, we proved the following statement.

Proposition $w^N(t)$ *converges weakly to* $\sqrt{(a^2+1)}w(t)$*.*

It looks promising to experiment with fitting *discrete time financial data* as a $TH_{a,b}$-model with variable parameter a estimated as a trigonometric series of some order to describe smooth volatility changes.

3.4.4 Asymptotic Normality for Additive Functions of m-MC Trajectories

Basically, our further limit theorems will be derived for finite m-MC by reducing them to the well-known case of MC on m-grams.

Rates of convergence in these theorems depend on the alphabet size which is significantly lower for sparse SCOT than for general m-MC.

The main steps of our AN straightforward derivation for ergodic m-MC as a corollary of that in [57] are:

- Given an ergodic m-MC $\{X_i\}$ with finite alphabet A, denote the induced 1-MC on m-grams (see Sect. 3.3.1) as $\{Y_i\}$. We *assume that MC* $\{Y_i\}$ *is ergodic*, which does not generally follow from ergodicity of X_N,
- This 1-MC $\{Y_N\}$ is a Harris invariant (see e.g. [57], Chap. 17) w.r.t. the induced probability distribution.

Let $g(\cdot)$ be a Borel function on $\mathbf{R}$.

- Define $f(Y_i) := f(X_i, X_{i-1}, \ldots, X_{i-m+1}) = \sum_{k=0}^{m-1} g(X_{i-k})$.
- Define $\bar{f}_N := (1/N) \sum_{i=1}^{N} f(Y_i), \bar{g}_N := (1/N) \sum_{i=1}^{N} g(X_i)$.
- If π is the stationary distribution and $E_\pi |f^2| < \infty$, then the ergodic theorem ([57], Sect. 17.3) guarantees that $\bar{f}_N \to E_\pi f$ with probability 1 as $N \to \infty$, and the central limit theorem holds for $\bar{f}_N$ ([57], Sect. 17.4):

$\sqrt{N}(\bar{f}_N - E_\pi f) \Rightarrow N(0, f_\pi^2)$ weakly, where $\sigma f_\pi^2 < \infty$ is the variance of f with respect to π.

- $(1/\sqrt{N} \sum_{i=1}^{N} f(Y_i) - E_\pi f) \Rightarrow N(0, \sigma f_\pi^2)$ weakly,
- $(1/\sqrt{N} \sum_(i = 1)^N \sum_{k=0}^{m-1} g(X_{i-k}) - E_\pi f) \Rightarrow N(0, \sigma f_\pi^2)$ weakly,
- $\sqrt{N}(m\bar{g}_N(X) - E_\pi f) \Rightarrow N(0, \sigma f_\pi^2)$ weakly.
- The joint convergence of the sample mean and sample variance to independent random variables is established for ergodic finite MC similarly. These limiting components have respectively the Normal and χ^2 distributions
- A sparse ergodic SCOT AN convergence rate under fixed context size is generally better than for the full m-MC. As far as we know, the case of proportional sample size and of the MC alphabet sizes is not yet studied.
- The above results justify t-distribution of our homogeneity test statistic of Sect. 3.5 based on studentized averages of SCOT log-likelihoods.

3.4.5 Asymptotic Expansion for Additive Functions

The asymptotic expansion for additive functions of MC trajectories appeared in [7, 8, 32, 40] under various conditions which include the case of a finite ergodic MC.

In [40], the *first terms* of asymptotic expansion under Cramer-type conditions are:

$$P(N^{-1/2}(\sum_{i=1}^{N} f(x_i) \leq x)) = \Phi_\sigma(x) + \phi_\sigma(x)q(x)N^{-1/2} + O(N^{-1}).$$

Here ϕ and Φ are respectively the PDF and the CDF of the central Normal RV with StD σ, where $q(\cdot)$ is expressed in terms of the Hermite polynomial.

Malinovsky finds an explicit expression for the polynomial $q(x)$.

This result can be generalized for m-MC by the method displayed above. We believe that the coefficient $q(x)$ for *sparse SCOT* is substantially less than that for general m-MC.

3.4.6 Nonparametric Homogeneity Test

We estimate the SCOT of the large stationary ergodic 'training' string T. Then, using the SCOT of T we first find the log-likelihood $L_Q(k)$ of query slices Q_k, and second, of strings S_k simulated from the training distribution of the same size as $Q_k, k = 1, \ldots, K$ (for constructing simulated strings, see algorithm in [39]).

We then find log-likelihoods $L_Q(k)$ of Q_k, $L_S(k)$ of S_k using the derived probability model of the training string, and the average $\bar{D}$ of their difference D.

Next, due to the asymptotic normality of the log-likelihood increments, we can compute the usual empirical variance V of $\bar{D}$ and the t-statistic t as the ratio $\bar{D}/\sqrt{V}$ with $K - 1$ degrees of freedom (DF). We find K^* from the condition that $t(K^*)$ is maximal. Then, the P-value of homogeneity is evaluated for the t-distribution with $K^* - 1$ DF.

3.4.7 Exponential Tails for Log-Likelihood Functions

Derivation of the exponential tails obtained in this section follows the pattern usual for the MC and is also somewhat similar to the bounds evaluation obtained earlier for the UC-based inference.

Generate an artificial N-sequence z^N independent of y^N and distributed as P_0 and denote by L_0 its log-likelihood given the SCOT model of the training string.

L is the query log-likelihood given the SCOT model of the training string.

Also assume that the joint distribution of S slices of size N converge to their product distribution in probability.

Theorem 3.4.2 *Suppose P_1, P_0 are SCOT, $D(P_1||P_0) > \lambda$ and we reject homogeneity if the 'conditional version of the Likelihood Ratio' test $\mathscr{T} = \overline{L} - \overline{L}_0 > \lambda$. Then* *the same error probability asymptotic as for LRT in the 'Stein' lemma is valid for this test.*

Proof Sketch Similar to That in Sect. 3.3.2 Under independent slices, their probabilities multiply. To transparently outline our ideas (with some abuse of notation) replace the condition under summation sign to a similar one for the whole query string: instead of $P_0(T' > 0) = \sum_{y^N, z^N : \bar{L} - \bar{L}_0 > \lambda} P_0(y)P_0(z)$, we write the condition under the summation as $L - L_0 > N\lambda$, which is approximated in probability under P_0 by $L_1(y^N) - L_0(z^N) > N\lambda$.

Thus

$$P_0(T' > 0) \leq \sum_{z^N} \sum_{y^N} : P_0(y^N) \leq 2^{-N\lambda - L(z^N)} P_0(y) P_0(z^N) \leq 2^{-N\lambda} \sum_{z^N} 2^{L_0(z^N)} = 2^{-N\lambda},$$

since $2^{L_0(z^N)}$ is a probability distribution. We refer to [78] for an accurate completion of our proof in a similar situation.

Informally again,

$$\lim_{N \to \infty} P_1(T' > 0) = \lim_{N \to \infty} P_1(N^{-1}(L_1(y^N) - L_0(z^N)) > \lambda) = D(P_1||P_0) + \varepsilon, \varepsilon > 0.$$

$|y|/N$ is in probability P_1 around $-\log P_0(y) = E_1(-\log(P_0(y))) + r$, $|z|/N$ is in probability P_0 around $-\log P_0(z^N) = h_0^N + r'$. As in the proof of Theorem 3.2.3, all the principal deterministic terms drop out, and we are left with the condition $r < \varepsilon + r'$. Its probability converges to 1 since both r, r' shrink to zero in the product (y, z)-probability as $N \to \infty$.

3.4.8 Application to the STI Analysis Under Colored Noise

Suppose as in [46] that we can assign an arbitrary t-tuple of binary inputs $\mathbf{x} := (x(j), j \in [t]), [t] := \{1, \ldots, t\}$, and measure the **noisy output** Z in a measurable space $\mathscr{Z}$ such that $P(z|y)$ is its conditional distribution given *'intermediate output'* $y = g(x(A))$ with a finite-valued function $g(\cdot)$, and $P(z|x(A))$ is their superposition (Multi-Access Channel (MAC)), where $x(A)$ is an s-tuple of 'Active Inputs' (as in [46]). Conversely, every MAC can be decomposed uniquely into such a superposition. We omit an obvious generalization to q-ary inputs.

The Mean Error probability (MEP) $\gamma > 0$ is the misidentifying s-tuple A probability of a test T based on a sequence of measurements $\mathbf{z} \in \mathscr{Z}^N$. This probability is extended over the direct product of uniform prior for A and noise distribution of $\mathbf{z}$.

It is wellknown that for **known MAC** $P(z|x(A))$, the Maximum Likelihood (ML)-decision minimizes the MEP for any design. It generalizes the Brute Force analysis of noiseless data for the case of known MAC distribution. If MAC is unknown, a

universal nonparametric test of computational complexity $O(t \log t)$ for IID noise and a random design is $\mathscr{I}$ [46].

The universal STI decision chooses maximal values of the $\mathscr{I}$ for input a and output, $a = 1, \ldots, t$. For an IID noise, $\mathscr{I}$ is the Empirical Shannon Information (ESI):

$$\mathscr{I}(\tau_N^N(A)) = \sum_{x(a)\in\mathscr{B}^{|A|}} \sum_{z\in\mathscr{Z}^{\mathscr{N}}} \tau(x(a), z) \log(\tau(z|x(a))/\tau(z)).$$

Let us now consider 'Colored' noise. Given arbitrary intermediate vector-output $\mathbf{y}$, the sequence $\mathbf{z}$'s conditional distribution is that of a *stationary ergodic SCOT random string* taking values from a finite alphabet $\mathscr{Z}^N$.

The lower bounds for the minimal sample size that provides a given MEP for colored noise can be derived with the 'entropy rate' $\lim_{N\to\infty}(I_\beta^{\mathbf{m}}(X_1^N(\lambda) \wedge \zeta_1^N)/N)$ instead of constant $I_\beta^{\mathbf{m}}(X(\lambda) \wedge \zeta)$. The limit exists due to the stationarity of the pair (IID X_i, ζ_i).

The STI-test for colored noise is defined as follows:

Define

$$\mathscr{U} = \mathscr{B} \times \mathscr{Z},$$

and consider, for a given $j = 1, \ldots, N$, two N-sequences with letters from $\mathscr{U}$:

$$u_j^N := (x_j(i), z(i)), i = 1 \ldots, N,$$

and

$$v_j^N := (x_j(i)(\times)z(i)), i = 1 \ldots, N,$$

taken from the original and generated product distributions; digitize them into binary sequences $\mathbf{U}_j^M, \mathbf{V}_j^M$ of appropriate length and evaluate the SCOT homogeneity statistic (see further) of the product P_0 and original distributions $P_1 = P_1^j$. The SCOT—trained log-likelihoods of these bi-variate strings are $\log P(\mathbf{V}^M) = L^M$—our main inference tool.

Consider a **query** binary sequence $\mathbf{U}^M$ distributed as P_1 and test whether the homogeneity hypothesis $P_0 = P_1$ contradicts the data or not. Let us extract from $\mathbf{y}^M$ several **slices** $\mathbf{U}_i, i = 1, \ldots, S$, of identical length N on a small distance between consecutive slices which is sufficient to ensure independence of slices for m-MC. Introduce strings $\mathbf{C}_i = (\mathbf{V}^M, \mathbf{U}_i)$. Define L^i—statistic as L^M before for slice $\mathbf{C}_i$ and $\overline{L}$ = average of all L^i. Similarly, $\overline{L}_0$ = average of all L^{0i} where $\mathbf{U}_i$ are replaced with independent P_0-distributed slices of the same length. Finally, the homogeneity test statistic is $\bar{R} = \overline{L} - \overline{L}_0$. The $\bar{R}$ test is shown in Theorem 3.4.2 to have the same exponential tails under P_0 as the asymptotically optimal Likelihood Ratio test if the error probability under an alternative is arbitrarily small but positive and fixed.

Our setting is natural if s = const, $t \to \infty$. This fact can be used to prove that the STI test is asymptotically optimal among all tests of affordable computational complexity $O(t \log t)$, based on separate influence comparison of inputs to the output of the system with stationary noise under a random design.

3.5 SCOT Homogeneity Test for Real Data

This section describes SCOT homogeneity test's statistical applications to real data from several fields.

We skip description of Rissanen's 'Context' algorithm implemented by M. Mächler and P. Bühlmann in 'Context' software [39] aimed at estimation of all **contexts** and corresponding transition probabilities. We note only that its restriction to a maximal alphabet size ≤ 27 forces, in particular, compression of our three-variate financial data into triples with three-valued components, which can make them less informative than the corresponding two-variate and uni-variate counterparts. The author of an alternative PST algorithm sent us its extension to larger alphabet size, which works much slower. The proof of consistency of 'Context' under exponential mixing of stationary data is in [24, 39, 63].

An elementary nonparametric CCC-test based on Conditional Complexity of Compression was previously applied to the authorship attribution of literary texts (LT) outlined in Sect. 3.6.

We use in Sect. 3.5 an alternative SCOT-based Likelihood Ratio (SCOTlr) nonparametric homogeneity test methodologically similar to the CCC-test. Its mathematical theory discussed in Sects. 3.3 and 3.4 for finite MCs is more elementary than that of the CCC-test.

We estimate the SCOT of the large stationary ergodic 'training' string T using the 'Context' software. This is then concatenated first with query slices Q_k and with strings S_k simulated from the training distribution of the same size as $Q_k, k = 1, \ldots, K$ (for constructing simulated strings, see algorithm in [39]).

We then find log-likelihoods $L_Q(k)$ of Q_k and $L_S(k)$ of S_k using the derived probability model of the training string and the average $\bar{D}$ of their difference D. Thus, the SCOT-approach difference from the CCC-methodology is in the **log-likelihood** direct evaluation instead of the **compressed increment lengths** evaluation. The latter only approximate log-likelihoods under certain conditions, including sufficient smallness of the slice size.

Next, due to the asymptotic normality of the log-likelihood increments, we can compute the usual empirical variance V of $\bar{D}$ and the t-statistic t as the ratio $\bar{D}/\sqrt{V}$ with $K - 1$ degrees of freedom (DF). We find K^* from the condition that $t(K^*)$ is maximal. Then, the P-value of homogeneity is evaluated for the t-distribution with $K^* - 1$ DF.

3.5.1 NASDAQ Data

We use historical Nasdaq data on multivariate daily returns for 498 days from April 4, 2011 to March 27, 2013 collected from *finance.yahoo.com* and converted into log returns (Fig. 3.8). We reduce the dimensionality of single-day returns via MATLAB version of Principal Component Analysis (PCA) and compress the data set to the sequence of one, two or three Principal Components (PC), displayed in Fig. 3.9 which describe a major part of the data variability. We fit their SCOT-stochastic model and apply it for discrimination between statistical properties of different parts of the data. We also compare our model with GARCH.

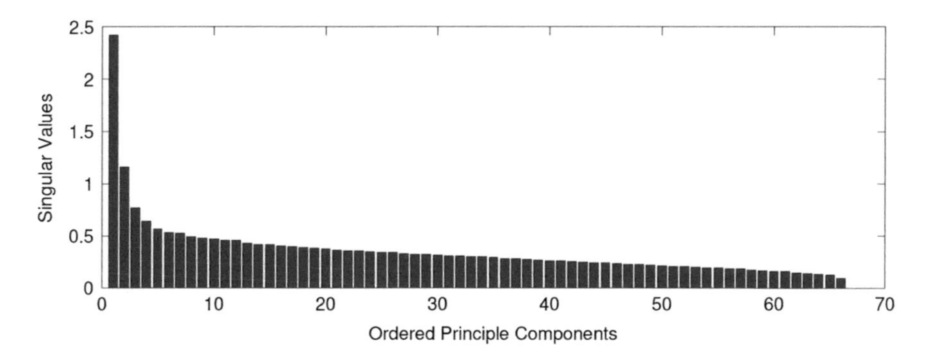

Fig. 3.8 Singular values of daily log-returns

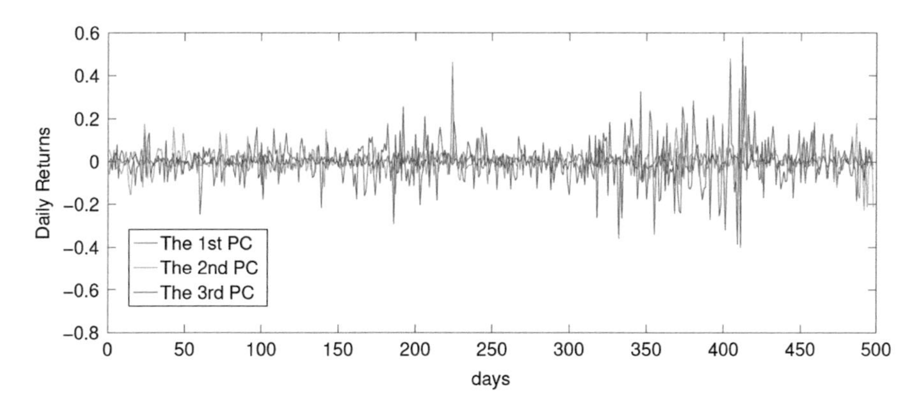

Fig. 3.9 The first three principal components of daily log-returns

3.5.2 Results for Three PC

The range of each PC is divided into three equal intervals (bins). Triples of PC-values are compressed to triples of integers from {1,2,3} according to their belonging to corresponding bins and their triples are labeled with 26 English letters from A to Z or the * symbol. The sequence of 498 triples is thus converted into a sequence of symbols, of which we display only a few:

NNNNNNENNNNNNNKKNNENNNEKNNKNNKNNNNNNNNENEN NNNENNNNNNNNNNNNNNNNNEEN....

Only eight of all 27 symbols were observed in the whole sequence, the maximal size of contexts turns out to be 2, and the number of leaves is five.

The detailed output of the Context program is displayed in [47], their part is Table 3.1.

The SCOT homogeneity t-test between 1–150 and 301–420 (quiet and volatile regions) is trained on 301–420 cut into 12 slices. The t-value is -2.53.

Inter-log-likelihood output has the mean $m_2 = -6.21$, variance $v_2 = 3.92$.

Intra-log-likelihood output has the mean $m_1 = -8.18$, variance $v_1 = 4.16$.

Prediction Accuracy

We estimate the Squared Bias to show the accuracy of our prediction by predicting 10 consecutive letters from the 131st to 140th, and for each letter, the prediction is based on training the preceding 130 letters. The variance of the log-likelihood of the simulated letters is 0.39. The Squared Bias is 0.048, which is 8.1 times less. This illustrates the adequacy of our prediction.

Also, we predict each of 20 letters of the quiet zone by training SCOT on 120 preceding letters. The coincidence rate of predicted and actual letters is about 75 %.

3.5.3 Results for Two PC

The range of each PC is divided into five equal bins. PC-values are compressed to only five integers, 1–5, according to their belonging to corresponding bins and their

Table 3.1 SCOT training result

Alphabet	'bdejkntw'
Number of alphabet	8
Number of letters	120
Maximal order of Markov chain	2
Context tree size	7
Number of leaves	5

pairs are labeled with 25 English letters from A to Y similarly to 3 PC case. Only eight of all 25 symbols were observed.

The detailed output of the Context program is displayed in [47], their part is Table 3.2. The SCOT homogeneity t-test between 1–150 and 301–420 (quiet and volatile regions) trained on 301–420 gives the t-value -2.84.

The Inter-log-likelihood output has the mean $m_2 = -11.08$, variance $v_2 = 4.59$.

The Intra-log-likelihood output has the mean m_1= -13.00, variance $v_1 = 1.81$.

3.5.4 Results for One PC

The Context program output is displayed in Table 3.3. The maximal size of contexts is three. The SCOT homogeneity t-test between 1–150 and 240–300 (two relatively quiet regions) trained on 1–150 gives the $t = -1.59$.

The Inter-log-likelihood output has the mean $m_2 = -10.88$, variance $v_2 = 1.67$.

The Intra-log-likelihood output has the mean $m_1 = -11.82$, variance $v_1 = 1.10$.

3.5.5 Prediction Accuracy

We estimate the Squared Bias to show the accuracy of our prediction by predicting 10 consecutive letters from the 141st to 150th (quiet zone), and for each letter the prediction is based on training the preceding 140 letters. The variance of the log-likelihood of the simulated letters is 0.07. The Squared Bias is 0.04.

Also, we predict each of 10 letters of the quiet zone by training SCOT on 140 preceding letters. The coincidence rate of predicted and actual letters was about

Table 3.2 SCOT training result

Alphabet	'bcghlmqr'
Number of alphabet	8
Number of letters	120
Maximal order of Markov chain	2
Context tree size	12
Number of leaves	8

Table 3.3 SCOT training result

Alphabet	'bcghlm'
Number of alphabet	6
Number of letters	150
Maximal order of Markov chain	3
Context tree size	6
Number of leaves	2

55 %. It is less than for the 3 PCs case partly because we must predict triples of bins rather than pairs of bins.

3.5.6 Follow-Up Analysis

One of the major **additional advantages of SCOTlr over CCC** is its more straightforward use for the follow-up estimation of contexts contributing the most to the discrimination between regions that were previously shown to be distinct.

For this aim, we propose cutting both the training and the query strings into several slices for estimating the mean frequencies and their empirical variances by their direct count, which approximates steady state probabilities and variances of contexts.

We obtain more transparent results of this follow-up analysis than those obtained with LZ-78 in [45].

The normalized frequency of occurrences of a context (with its frequency exceeding a threshold) of size n is Asymptotically Normal (AN) with variance σ_i^2, σ_i can be estimated via these frequencies in slices; $i = 0$ for the training string and 1 for query. The normalized difference between frequencies for 0 and 1 cases is then AN with variance as sum of the above variances.

For every SCOT context of the whole Training text, one can evaluate its multiplicities in $k(Tr)$ Training and $k(Qu)$ Query slices of the SAME LENGTH, their corresponding empirical means $m(Tr)$, $m(Qu)$, and empirical variances $V(Tr)$, $V(Qu)$. The usual t-statistics are

$$T(k) = (m(Tr) - m(Qu))/\sqrt{(V(Tr)/k(Tr) + V(Qu)/k(Qu))}. \tag{3.7}$$

Choose $k(Tr)$ and $k(Qu)$ from the condition that $t(k^*(Tr), k^*(Qu))$ is maximal and slice sizes are equal. The slice size must be several (say, not less than 5) sizes of the context considered. We then order t^* for different contexts starting from their maximal absolute values.

We exclude the contexts with frequencies less than a threshold in our first applications, in spite of the possibility of losing some unstable but potentially significant information contained in rare contexts.

Thus, one can find p-values for their equality and order these P-values starting with the minimal ones.

In the Two PC case, the quiet region has the patterns L (indicating that the first PC-value is located in the second bin and the second PC-value is located in the third bin) and H (indicating that the first PC-value is located in the third bin and the second PC-value is located in the second bin) while the volatile region has the pattern B (indicating that the first PC-value is located in the first bin and the second PC-value is located in the second bin) and Q (indicating that the first PC-value is located in the fourth bin while the second PC-value is located in the second bin).

In the Three PC case, the quiet region has the pattern N (indicating that all three PC-values are located in the second bin) while the volatile region has the pattern E (indicating that the first PC-value is located in the first bin while the second and third PC-values are located in the second bin).

3.5.7 Comparison with GARCH

In this subsection, we make a comparison between SCOTlr-discrimination and that of GARCH model [25, 28] applied to two different sets of financial data.

The first data set we use is the daily log-return data of Apple Inc. starting from January 2, 2009 (Fig. 3.10). By observation, we pick the volatile region (the first 450 days returns) and the quiet region (the 500th to 600th days returns) to make a comparison. We first fit the data with the GARCH(1,1) modeled using the

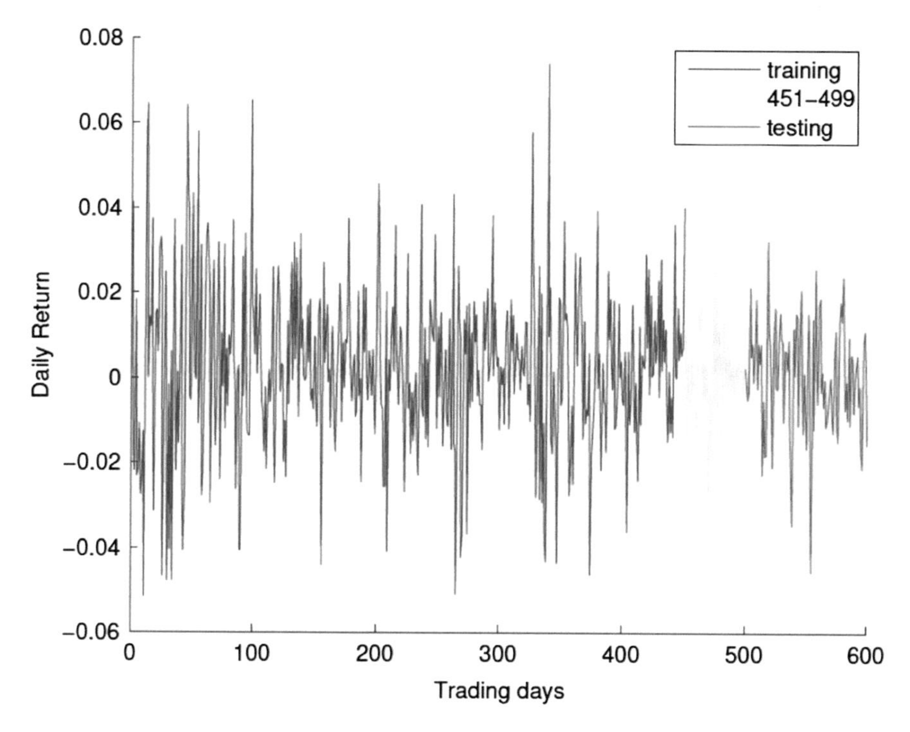

Fig. 3.10 Apple daily returns

MATLAB(R2011a) GARCH toolbox.

$$y_t = C + \epsilon_t \tag{3.8}$$

$$\epsilon_t = \sigma_t z_t \tag{3.9}$$

$$\sigma_t^2 = \kappa + G_1 \sigma_{t-1}^2 + A_1 \epsilon_{t-1}^2 \tag{3.10}$$

Let $\hat{\alpha}_1$ and $\hat{\beta}_1$ be the estimators for GARCH(1) and ARCH(1) in the first model. Similar notation can be defined for $\hat{\alpha}_2$ and $\hat{\beta}_2$. From the results, we have $z_1 = \frac{\hat{\alpha}_1 - \hat{\alpha}_2}{\sqrt{\sigma_{\hat{\alpha}_1}^2 + \sigma_{\hat{\alpha}_2}^2}} \doteq 2.16$ and $z_2 = \frac{\hat{\beta}_1 - \hat{\beta}_2}{\sqrt{\sigma_{\hat{\beta}_1}^2 + \sigma_{\hat{\beta}_2}^2}} \doteq -1.70$. The P-values obtained are $P_1 = 0.03$ and $P_2 = 0.09$.

We apply SCOTlr to the same data. The homogeneity t-test between 1–450 and 500–600 (quiet and volatile regions) trained on 1–450 shows that the t-value is -16.02. Thus, the P-value $P < 0.00001$. This P-value by SCOTlr is much smaller than the Z-score by GARCH.

We also use the **first** principal component of Nasdaq daily log-return data for comparison with GARCH. Again, let $\hat{\alpha}_1$ and $\hat{\beta}_1$ be the estimators for GARCH(1) and ARCH(1) in the first model. Similar notation can be defined for $\hat{\alpha}_2$ and $\hat{\beta}_2$. From the results, we have $z_1 = \frac{\hat{\alpha}_1 - \hat{\alpha}_2}{\sqrt{\sigma_{\hat{\alpha}_1}^2 + \sigma_{\hat{\alpha}_2}^2}} \doteq -1.18$ and $z_2 = \frac{\hat{\beta}_1 - \hat{\beta}_2}{\sqrt{\sigma_{\hat{\beta}_1}^2 + \sigma_{\hat{\beta}_2}^2}} \doteq 2.16$. And thus, the P-values are $P_1 = 0.24$ and $P_2 = 0.03$.

We divide the range of the first PC of Nasdaq daily log-returns into 27 bins. Each bin is labeled with 26 English letters from A to Z and symbol *. The sequence of the first PC of daily log-returns is converted into a sequence of symbols. The homogeneity t-test between 1–150 and 301–420 (quiet and volatile regions) trained on 301–420 shows that the t-value is -7.05. Thus, the P-value is $P < 0.000001$. This P-value obtained by SCOTlr is drastically smaller than the P-value by GARCH.

3.5.8 GARCH and Epicycles

Our SCOT models from Sect. 3.4 show that memory can become unnoticeable under continuous time limit. Modeling volatility can be done within simple SCOT models before taking continuous time limit. GARCH artificial memory modeling of volatility in the memoryless background resembles the pre-Copernicus view of the world. In the Ptolemaic world (Fig. 3.11), the planets are assumed to move in a small circle called an epicycle, which in turn moves along a larger circle called a deferent. Both circles rotate clockwise and are roughly parallel to the plane of the sun's orbit (ecliptic). Despite the fact that the system is considered geocentric, the planets' motion was not centered on the earth but on what is called the eccentric. The orbits of planets in this system are epitrochoids. The epicycle rotated and revolved along the deferent with uniform motion.

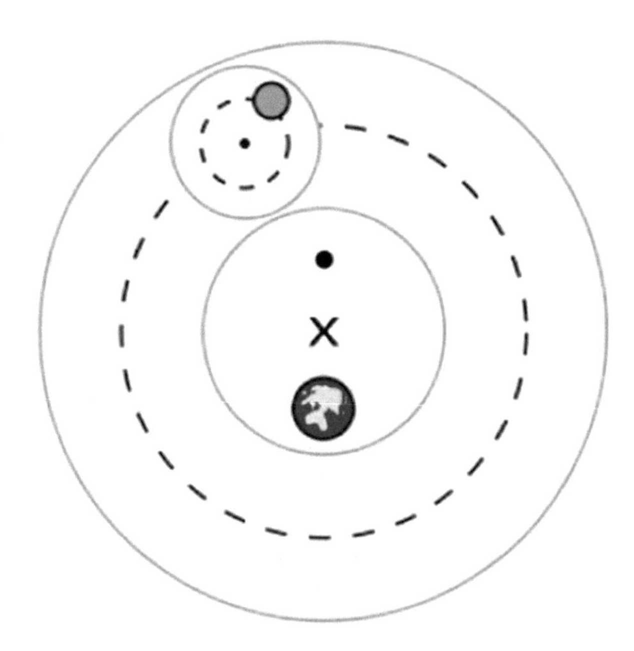

Fig. 3.11 Ptolemaic world

3.5.9 *Madison vs. Hamilton Discrimination of Styles*

The Federalist papers written by Alexander Hamilton, John Jay and James Madison appeared in newspapers in October 1787–August 1788. Their intent was to persuade the citizens of the State of New York to ratify the US Constitution. Seventy-seven essays first appeared in several different newspapers all based in New York and then eight additional articles written by Hamilton on the same subject were published in booklet form. Ever since that time, the consensus has been that John Jay was the sole author of five (Nos. 2–5, No. 64) out of a total 85 papers, that Hamilton was the sole author of 51 papers (Hf), that Madison was the sole author of 14 papers (Mf, Nos. 10, 14, 37–48) and that Madison and Hamilton collaborated on another three (No. 18-20). The authorship of the remaining 12 papers (Df, Nos. 49–58, 62, 63) has been in dispute; these papers are usually referred to as the disputed papers. It has been generally agreed that the Df papers were written by either Madison or Hamilton, without consensus on particulars. On the other hand, [42, 60, 77] and other stylometry attributors have given all Dfs to Madison. Our goal has been to answer the following two questions:

1. Does the SCOT-methodology attribute all Mfs to Madison and reject significantly the identity of the Hf style with that of Mfs?
2. What contexts are most statistically different between Mf and Hf?

   The answer we have obtained is 'yes' for the first question: Mf is attributed to Madison; identity of styles in Hf and Mf is rejected.

2. Our goal is to find patterns that occur in one collection significantly more often than in the other. The patterns (SCOT contexts) we will examine are those that arise during the SCOT training.

To generate meaningful statistics, we divide each file into *slices* of a given size, and run the ‘Context’ program to find the log-likelihood of each slice. We then apply the t-test to find out whether the styles of articles written by the two authors are significantly different. For each pattern, we calculate the mean number of occurrences for each file. Finally, we get the t-value for each pattern (SCOT context).

Here is a detailed description of the algorithm:

1. For each of the two files Madison (M) Hamilton (H):

   a. Train the SCOT model on Madison’s text.
   b. Cut both the M and H texts into slices of equal given size.
   c. For each slice:

      i. Use Madison SCOT to calculate the log-likelihood of each slice.
      ii. Calculate the mean value, the variance of slices of M and H respectively.
      iii. Calculate the t-value as $\frac{|L_1-L_2|}{\sqrt{v_1/n_1+v_2/n_2}}$. where L_1 and L_2 are the mean values of log-likelihood of slices of M and H, v_1 and v_2 are variances over slices of M and H, and n_1 and n_2 are the numbers of slices of M and H respectively.

   d. For each pattern (SCOT context):

      i. Calculate the empirical mean occurrence.
      ii. Calculate the empirical variance of occurrence.
      iii. Calculate the t-value as $\frac{|m_1-m_2|}{\sqrt{v_1/n_1+v_2/n_2}}$, where m_1 and m_2 are the means of occurrence number of one pattern in each slice of M and H, and v_1 and v_2 are occurrence variances of M and H; n_1 and n_2 are the numbers of slices of M and H respectively,

Tables 3.4 and 3.5 show detail of the data input and output.

Table 3.4 Data from Madison

fileproc10.txt	fileproc14.txt	fileproc37.txt	fileproc38.txt
fileproc39.txt	fileproc40.txt	fileproc41.txt	fileproc42.txt
fileproc43.txt	fileproc44.txt	fileproc45.txt	fileproc46.txt
fileproc47.txt	fileproc48.txt		

Table 3.5 Data from Hamilton

fileproc1.txt	fileproc6.txt	fileproc7.txt	fileproc8.txt
fileproc9.txt	fileproc11.txt	fileproc12.txt	fileproc13.txt
fileproc15.txt	fileproc16.txt	fileproc17.txt	fileproc31.txt
fileproc32.txt	fileproc34.txt		

We first combine all of 14 Madison's articles into one file and use it as the training data. The maximal memory n is set to be 15. (Thus, at most 15 *letters* predict the next letter.)

The total number of letters from Madison's articles is 228,744 (Table 3.6). We cut the whole Madison data into nine slices and each slice contains 25,416 letters.

To perform an intra-SCOT t-test we use each slice of Madison's data as a query string and use the remaining eight slices as a training string. We compute the log-likelihoods of each query string (Table 3.7).

Then we import the data from Hamilton. The total number of letters of Hamilton's articles is 152,496. We cut the letters into six slices so that each slice contain 25,416 letters ($25416 \times 6 = 152496$).

To perform an inter-SCOT t-test we use the total training result of Madison to evaluate the log-likelihood of each slice of Hamilton (Table 3.8).

Plugging in the numbers into the following formula for the t-test

$$t = \frac{ł_1 - ł_2}{\sqrt{var_1/n_1 + var_2/n_2}},$$

we get the t-value 2.690809.

Similar formulas give the insignificant t-value -0.7864356 for "Inter- and Intra-comparison between Mf and itself (displayed in Tables 3.7–3.9).

Table 3.6 SCOT training results

Alphabet	'*abcdefghijklmno pqrstuvwxyz'
Number of alphabet	27
Number of letters	228,744
Maximal order of Markov chain	13
Context tree size	3365
Number of leaves	2353

Table 3.7 Inter-log-likelihood output

−27863.56	−28047.04	−27236.75	−26559.74
−24995.70	−27173.49	−26209.20	−25182.81
−25622.52			
	Mean value $m_2 = -26543.42$		Variance $v_2 = 1261004$

Table 3.8 Intra-log-likelihood output

−28552.20	−28462.64	−28511.57	−28234.03
−27227.31	−26510.97		
	Mean value $m_1 = -27916.45$		Variance $v_1 = 721562.6$

Table 3.9 Inter-log-likelihood output

−27725.26	−27341.10	−26948.34	−26352.83
−23933.95	−26703.40	−25770.18	−24609.77
−25525.76			
	Mean $m_2 = 26101.18$		Variance $v_2 = 1585075$

The SCOT model in this application is much more complicated than that for the financial data set: the maximal context size turns out to be 13, number of leaves is 2353. Details are displayed in the table-output of the 'Context' program.

Cutting Madison's text into 14 slices and performing a t-test between Madison's and Hamilton's data by the method described above, we get the t-score 3.10.

Cutting the Madison's text into 20 slices and performing a t-test between Madison's and Hamilton's data by the method shown above, we get the t-score 4.02.

Frequencies of SCOT Contexts

We have calculated frequencies of each SCOT context over slices and found patterns with the most significantly different frequencies between Madison's and Hamilton's articles. This mining might be of interest for linguists.

The majority of differentially expressed patterns found in [45] using LZ-78 turned out to be in our larger list.

1 The following are patterns that Madison used significantly more frequently than Hamilton (we cut Madison's data into 14 slices, **symbol * denotes space**):

For P-value < 0.005:

nt*, rs*, fore*, han*, de*, ed*b, by, by*, by*t, f*, nd*be, orm, *on*th, *th, *on*t, ese*, i*, over*, ay*be, und*, d*, he*n , g*the*, if*, der*, he*, ple*, ay*b, he*for, he*or, both, c*, re*, *ele, bot, ix, lt*, sume, ho*, ke*, *el, y*the*, latt, latte, oin*, *bo, *ne, by*o, veral*, *han, seco, *lat, d*th, in*, tas, ore*, eside, mer*, is*, ewe, rs*a, *se, tes*, ns*, at*th, lf* , d*on, ver*.

For $0.005 \leq P - \text{value} < 0.01$:

ca*, ose*, lst, *de, vera, *we, e*, nve , *cri, ere*, dur, sever, he*se, pe*, at*t, tuti, e*g, nsti, titu, w*f, nta, he*f, *tit, nin, sing, tit, ution*, rif.

2 The following are patterns that Hamilton used significantly more frequently than Madison:

For P-value < 0.005:

upo, pon, *k, *up, duc, ne, uis, ance, *le, om, om*, nati, wou, ne*, *at, *at*, ct, *wo, nation, id, way, *nat, ould*, e*ar, ontr, ity*, *ther, tim, qu, there, there*, *to, *to*, ut, ract, ign, e*to*, ndu, ont, ies, lit, eso, *if, erac, eng, it, it*, thin, wea, oul, dic, es*of*, *this*, ld, ld*, ilit, kind, *us, *us*, va.

For $.005 \leq P - \text{value} < 0.01$:

he*i, ten, arc, rri, y*t, lig, ig, s, s*, ation, sit, *lan, comm, ld*b, time, uisi, ces, iv.

We also performed this analysis by cutting data into nine and 20 slices. The results are available by request.

3.5.10 Helium Emissions and Seismic Events

Complex processes in the earth's crust preceding strong earthquakes imply changes of concentrations of certain chemicals dissolved in ground water samples from deep wells. Of particular interest in the region of Armenia studied in [31] are the Helium emissions into the water samples from deep wells (unlike studies in [62] of near-surface emissions) that appear attractive as a potential earthquake predictor.

We consider an approximately 10-year-long set of Helium emissions data during the 1980s and 1990s from three deep Armenian wells in Kadaran, Ararat and Surenavan. The earthquake dates in their vicinity shown in Fig. 3.12 were sent to us by Dr. E.A. Haroutunian (Inst. for Informatics and Automat. Problems, Armenian Acad. Sci.) for our robust analysis. In [31] they showed separately for each well that the Wilcoxon statistical test distinguishes between the quiet region of the plot and that preceding strong earthquakes. The Wilcoxon test was derived under assumption of sample independence, which does not hold in this application. Our problem was to check if VLMClr can distinguish between the regions given above. Instead of separate studies of data from the three wells we used PCA-compressed data. After the observations started, the earthquake days were 529; 925; 1437; 1797; 1997; 2470; 2629 and 2854. The singular value plot (Fig. 3.13) suggests using either one or two PCs.

In the one-PC case, the range is divided into 27 equal bins; PC-values are labeled with 26 English letters from A to Z and symbol * according to their belonging to bins. The 'Context' gave us the following parameters of the stochastic context tree shown in Table 3.10.

The quiet region 1–400 has the letters "hijklmnopqr", while the volatile region before and after an earthquake, 429–568, has the letters "opqrstuvwxy". It is impossible to train either region to predict the other one. It is also not necessary to do that because one can easily distinguish between different regions by observing

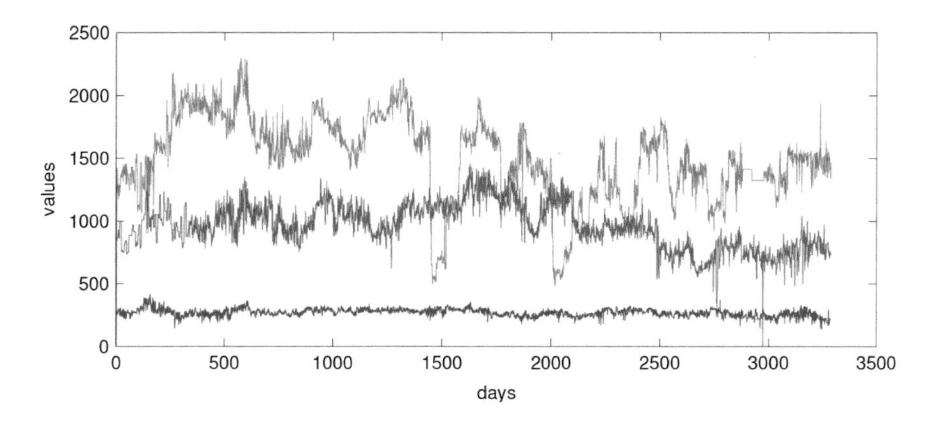

Fig. 3.12 Helium emissions data from three deep Armenian wells

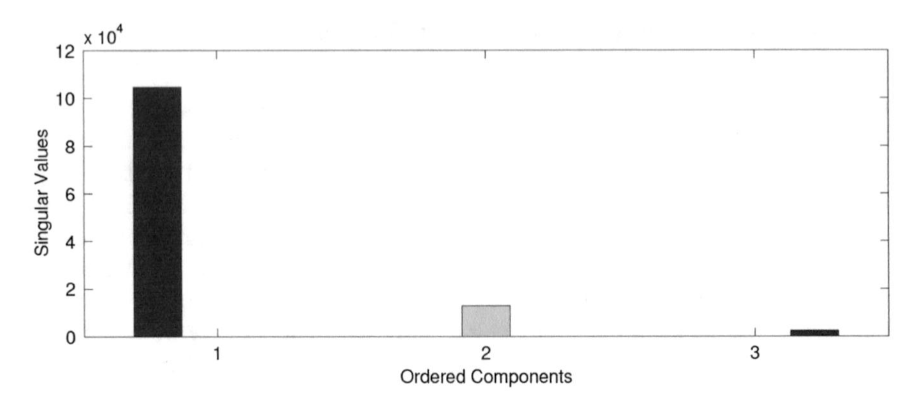

Fig. 3.13 Singular values

Table 3.10 SCOT training result

Alphabet	'abfghlmnqrstw'
Number of alphabet	17
Number of letters	400
Maximal order of Markov chain	3
Context tree size	28
Number of leaves	21

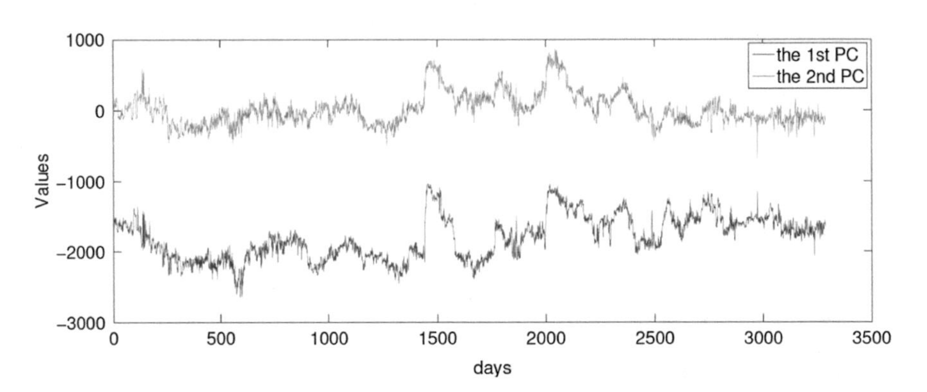

Fig. 3.14 Top two PCs

that the quiet region has the first PC-value up above a level "n" (corresponding to the value −1815.101), and the volatile region has the first PC-value down below a level "s" (corresponding to the number −2174.363).

In the Two PCs case (Fig. 3.14), the range of each PC is divided into five equal bins. PC-values are compressed to five integers according to their belonging to bins and their pairs are labeled with 25 English letters from A to Y. We list the first few letters as an example of labeled letters:

r r r r r r r r r r r r r r r r r r m m m m q q q q q q q r q q q q q r q q r r m m r m r r r q q q q q q q q r r r r r r r r g g g...

Table 3.11 SCOT training result

Alphabet	'abfghlmnqrstw'
Number of alphabet	13
Number of letters	400
Maximal order of Markov chain	3
Context tree size	22
Number of leaves	15

Training the quiet region between 1–400 will provide the training result shown in Table 3.11.

The homogeneity t-value between 1–400 (quiet region) and 429–528 (before earthquake) is 4.4, which means that these two regions are quite different. By calculating the t-value for each context, we get the context that distinguishes the most between these two regions. 'l' (the first PC value in the third bin and the second PC value in the second bin) is the typical pattern of volatile regions before an earthquake and 'b' (the first PC value is in the first quartile and the second PC value is in the second quartile) is the typical pattern of quiet regions.

The homogeneity t-test between 1–400 (quiet region) and 529–568 (*after* earthquake) is 0.8, which means that we find not much difference between the quiet region and the region after earthquake. In addition, we find an interesting letter 'c' (it means that the first PC is located in the first bin and the second PC is in the third bin), which can be an indicator for quiet times to follow because, when each 'c' appears, there were at least 100 quiet days beyond it in the future.

3.5.11 Discussion and Conclusions

Our results show that the SCOT modeling can be useful in various prediction and inference problems of certain time series.

Our next goal will be to use alternatives to the 'Context' program T for enabling processing with larger alphabets faster using parallel computing and running online versions of the SCOT construction for the fast online detection of the abrupt changes in data statistical profiles and for processing musical scores.

3.6 UC-Based Authorship Attribution

We study a new *context-free* computationally simple stylometry-based attributor: the *sliced conditional compression complexity (sCCC or simply CCC) of literary texts* introduced in [41] and inspired by the incomputable Kolmogorov conditional complexity. Other stylometry tools can occasionally almost coincide for different authors. Our CCC-attributor is asymptotically strictly minimal for the true author, if the query texts are sufficiently large but much less than the training texts, the universal compressor is good and sampling bias is avoided. This classifier simplifies the homogeneity test in [68] (partly based on compression) **under insignificant difference assumption of unconditional complexities of training and query texts**. This assumption is verified via its asymptotic normality [75] for IID and Markov sources and normal plots for real literary texts. It is *consistent* under large text approximation as a *stationary ergodic sequence* due to the *lower bound for the minimax compression redundancy of piecewise stationary strings* [56] (see also our elementary combinatorial arguments and simulation for IID sources). The CCC is based on the *t-ratio* measuring how many standard deviations are in the mean difference of the slices' CCC, which enables evaluation of the corresponding P-value of statistical significance. This is based on *asymptotic normality of the slices' CCC which is empirically verified by their normal plots constructed in all cases studied* and theoretically discussed in Sect. 3.2.3.

The *asymptotic CCC study in Sect. 3.2.3 is complemented by many literary case studies*—attribution of the Federalist papers agreeing with previous results; significant (beyond any doubt) mean CCC-difference between two translations of Shakespeare sonnets into Russian, between the two parts of M. Sholokhov's early short novel and less so between the two Isaiah books from the Bible; intriguing CCC-relations between certain Elizabethan poems. At the same time, two different novels *deliberately written in different styles* by S. Brodsky's and all Madison's Federalist papers showed insignificant mean CCC-difference.

Another application of universal compressor-based statistical methodology for screening out sparse active inputs in general systems disturbed by stationary ergodic noise with memory is outlined in [46].

Consider $U(Q,A) = |S_c| - |A_c| - |Q|$. Quantity $U(Q,A)$ *mimics the Ryabko and Astola statistic* ρ. In $U(Q,A)$ we *replace their empirical Shannon entropy* h^* of the concatenated sample S (based on n-MC approximation) with $|S_c|$ since both are asymptotically equivalent to $h(|Q| + |A|)$ for identical distribution in Q, A with entropy rate h and exceed this quantity for different A, Q .

Test ρ is asymptotically invariant w.r.t. interchanging A, Q and *strictly positive* for *different* laws of A, Q if $a < |A|/|Q| < 1/a, a > 0$). The last but not the first property would seem to hold also for $U(Q,A)$ in some range of $|A|/|Q|$ due to the **lower bound for the minimax mean UC-compression redundancy of piecewise-stationary sources** [56], which is logarithmic in $(|Q| + |A|)$. It is proved in [26] that the sliding window size is of order $\log(|Q| + |A|)$ in a version LZY of the LZ-compressors which is UC. Also, [71] constructed a UC with the 'redundancy price for one jump in piecewise stationary regime' asymptotically equivalent to the above-mentioned Merhav's lower bound. Moreover, Ryabko (personal communication) argues that **every UC can be modified in such a way that the FAUC property holds**:

FAUC (Fast Adapting UC) For any given SEDs P_1, P_2 with equal entropy rates, cross-entropy $D(P_1||P_2) > \epsilon > 0$, A, Q distributed as P_1, Q' distributed as P_2, independent $A, Q, Q', |Q| = |Q'|$ and any $b > 0$, it holds

$$\mathbf{E}[|(A, Q')_c| - |(A, Q)_c|] = o(|Q|^b).$$

In what follows, we can assume that we deal with a FAUC. Let us emphasize that the FAUC condition is different from (although related to) the small mean redundancy property for SED.

Conjecture Popular UC used in compressing LT are FAUC.

Claim The U performance on IID extensive simulations in a large range of $|Q|$ (made several years ago by Ph.D. student of Northeastern University Stefan Savev) was not as predicted above (actually, empirical mean of U was negative) due apparently to the additional subtracting of $|Q_c|$. For small $|Q_c|$ this is due to unexpected large 'adaptation value' of $|Q_c|$, since 'entropy' asymptotics is not yet attained. For large $|Q_c|$, the small increase of U due to inhomogeneity 'is drowned' in the large noise of variable $|S_c|$. Averaging different slices of identically distributed moderately large $Q_i, i = 1, \ldots,$ can make mean U positive, yet this does not seem appropriate in our applications.

3.6.1 CCC- and CC-Statistics

Fortunately, another statistic, CCC, defined below, overcomes the shortfalls of statistic U.

In our applications, $|A|/|Q|$ is very large to statistically assess reliability of attribution and bounded from above by an approximate empirical condition $|Q| \geq$ 2000 bytes (requiring further study) for appropriateness of SED approximation. The *Conditional Complexity of Compression* of text B given text A is defined as

$$CCC(Q|A) = |S_c| - |A_c|, \tag{3.11}$$

$$CCC(B|A) = |S_c| - |A_c|, CCCr(B|A) = CCC(B|A)/|B|. \tag{3.12}$$

The CCC mimics an abstract conditional Kolmogorov Complexity in our settings and measures how adapting to patterns in the training text helps to compress the query text.

Introduce an empirically centered version $CCC'(B|A)$ by extracting $CCC(B|A')$ from $CCC(B|A)$, where $A' : |A'| = |A|$ is generated by the same SED as A.

Claim $CCC'(B|A)$ approximates the most powerful Likelihood Ratio Test of Q, A homogeneity under our condition on sample sizes and validity of SED approximation for both Q, A

Proof is sketched in Sect. 3.2.3.

The only difference of CCC from U is canceling the $|Q_c|$ removal, which prevents inhomogeneity increase in S_c from showing up because of the larger than expected 'adaptation' value of Q_c.

In our case studies, we average sliced CCC of text $Q_i, i = 1, \ldots, m = [|Q|/L]$, given the firmly attributed text A, dividing the *query text* Q into slices of equal length L. Universal compressors used are the same for all sizes of texts.

$$\overline{CCC(Q|A)} := \sum_{i=1}^{m} \frac{CCC(Q_i|A)}{m}, CC(Q_i) = |Q_i|, \tag{3.13}$$

$$\overline{CC(Q)} := \sum_{i=1}^{m} \frac{CC(Q_i)}{m}, \tag{3.14}$$

$$\sigma(\overline{CCC}) = \sqrt{\sum_{i=1}^{m}(CCC(Q_i|A) - \overline{CCC(Q|A)})^2/m(m-1)}, \tag{3.15}$$

$$\sigma(\overline{CC}) := \sqrt{\sum_{i=1}^{m}(CC(Q_i) - \overline{CC(Q)})^2/m(m-1)}, \tag{3.16}$$

$$\overline{\sigma}(Q, Q', \kappa) := \sqrt{\sigma^2(Q, \kappa) + \sigma^2(Q', \kappa)}, \tag{3.17}$$

where κ denotes either CC or CCC.

We call the first two empirical quantities '*Mean CCC(Q)*' *and* '*Mean CC(Q)*' respectively.

Introduce t-statistics for independent Q, Q':

$$t(CCC(Q, Q'|A)) := |\overline{CCC(Q|A)} - \overline{CCC(Q'|A)}|/\overline{\sigma}(Q, Q', CCC), \tag{3.18}$$

$$t(CC(Q, Q') := |\overline{CC(Q)} - \overline{CC(Q')}|/\overline{\sigma}(Q, Q', CC). \tag{3.19}$$

CC-statistics for Q, A are called insignificantly different at some significance level if the corresponding t does not exceed its critical value (which practically is chosen around 1.5–2).

Claim Both our case studies and the statistical simulation show that the sliced CCC-approach has a good homogeneity discrimination power in this range for moderate $|Q|$) and much larger $|A|$ in a surprisingly wide range of case studies with **insignificantly varying mean unconditional complexity** CC of compression.

3.6.2 CCC-Sample Size Requirements

Statistical testing of the latter condition is straightforward due to the **asymptotic normality** results of the compression complexity for IID and MC sources described in [75] and normal plots for LT.

The very technical indirect proof in [75] on around 50 pages and a direct probabilistic proof in [1] on only 30 pages (valid only for the symmetric binary IID case) have an extensions for CCC which theoretically supports a quite **unusual sample size relation** for the UC-homogeneity testing: **sample size of the training text must dramatically exceed those of slices of a query text**. We present our proofs in Sects. 3.2.2–3.2.3 based on natural assumptions about UC.

Informal Justification The training test A being fixed, $VarCCC(Q_i|A)$ of independent copies $Q_i, i = 1, \ldots, N$, of the query text Q are of order of $|Q|$ due to almost renewal-type pattern acquiring and the practically finite mean memory size of UC, while the mean increase in $CCC(Q|A)$ redundancy for *different distributions* of Q and A *as compared to their identity* is $o(|(A|Q)|^b)$ for any $b > 0$ for a FAUC. An accurate upper bound for adaptation period even for LZ78 is absent so far. Perhaps, they can be obtained extending LZ-78 upper bounds for redundancy in [69].

Thus, the t-ratio is negligible under the asymptotics $|A| \to \infty, 0 < \epsilon < |Q|/|A|$. Malyutov [41] explains this informally as follows: if the training A and alternative style query text Q sizes are comparable, then two flaws in homogeneity testing happen. First, a UC adapts to both at the extra length cost $o(|(A|Q)|^b)$ for any $b > 0$; this extra amount of $CCC(Q|A)$ is hidden in the noise with $VarCCC((A|Q)|$ of order $|(A|Q)|$. Second, the mean $CCC(Q|A)$ of larger slices of query texts have a **larger bias** due to **self-adapting of UC to the slices' patterns**.

This makes sample size requirements and symmetry arguments in [14] unappealing. [14] is also based on the conditional compression complexity although it **ignores the assessment of the statistical stability**. This explains examples of [14] misclassification shown in [65]. It can explain also the roots of early heated discussion around simpler development in [5], where the *sample size relation and statistical stability* issues were not addressed.

3.6.3 Naive Explanation of CCC-Consistency on Toy Example

We study here performance of the CCC-attributor in two ways:

1. justify it for very distinct distributions of query and training IID sequences and
2. show the results of simulation when these distributions can be closer. Consider a training binary Bernoulli(1/100) sequence $X_1^{10000000}$ with $P(X = 1) = p = 1/100$ and the query Bernoulli(0.99) sequence Y_1^{1000} with the opposite distribution $P(Y = 0) = 1/100$ and compare the lengths of LZ-78-compressed sequences $X_1^{1001000}$ and $X_1^{10000000}Y_1^{1000}$. Note that the entropies h of X and Y are

the same and thus both belong to $M(h), h = -p \log p - (1-p) \log p$. Let us support discussion of asymptotic performance of CCC in Sect. 3.6.1 by direct arguments.

The classical von Mises's results state that the number of a pattern 'r repeated 1s' in a Bernoulli(p) sequence of length N has the Poisson(λ) distribution if $Np^r(1-p) = \lambda$ for large N, (see [19], problem 11.26. The cardinality of patterns is understood there in a slightly different sense which does not influence our argument significantly).

Thus $X_1^{10000000}$ contains only the Poisson(1) distributed number of 111-patterns (i.e. only one such pattern in the mean) and *much less likely patterns with a larger number of* 1*s*. The additional length of compressed $X_1^{1001000}$ w.r.t. the length of compressed $X_1^{10000000}$ is due most likely to few occurrences of *large size patterns consisting mostly of* 0*s* in the continuation of the sequence.

The length of the LZ-compressed file is approximately $c \log c$ bits, where c is the number of distinct patterns in the initial string. The concatenated sequence $X_1^{10000000} Y_1^{1000}$ contains most likely **more than hundred new patterns** w.r.t. $X_1^{10000000}$ *consisting mostly of* 1*s*, and thus the compressed $X_1^{10000000} Y_1^{1000}$ contains hundreds of additional bits w.r.t. compressed $X_1^{1001000}$ most likely.

Remark Von Mises (see [19], Sect. 13.7) and [75] prove the asymptotic normality of the patterns' cardinality in Bernoulli sequences, which agrees with our empirical CCC-normality plots.

A MATLAB simulation (with the code written by D. Malioutov (an MIT student at that time) using the commercial update of LZ-78 for UNIX systems) compared the CCC of I.I.D. binary query strings of length N_2, generated first for the same randomization parameter p_1 as that for the training string of length N_1, and CCC for the second query string with the complementary randomization parameter $p_2 = 1 - p_1$ (*having the same unconditional CC*) in a wider range of p_1 than that in our toy example. See tables in [42].

3.6.4 Methodology

Firmly attributed corpora are usually referred to as *training texts* for training the compressor and the text under investigation is referred to as the *query text. Query texts* may be disputed ones or those used for estimating the performance of attributors.

Equally sized slices Q_i, T_i of *query and training texts* $Q, T(k)$ are used for several slice sizes. We calculate the averages over slices of $\overline{CC(T(k))}, \overline{CCC(Q|T(k))}$ and their empirical standard deviations for each training text $T(k)$ to analyze authorship with the CCC-analysis. Comparing $\overline{CCC(T(k)|T)}$ of few *query texts* is also used sometimes, keeping the *training text* T fixed.

$\overline{CCC(Q(k)|T)}$ may be interpreted as empirical generalized distance similarly to the cross-entropy which it presumably approximates. Thus we can compare CCC of various training texts for a fixed query text and vice versa. This is as good as for the cross-entropy, which is the main asymptotic tool of statistical discrimination.

Equality of slice lengths makes comparison of CCC and CC equivalent to comparing their corresponding normalized versions,

$$CCr = |A_c|/|A| \tag{3.20}$$

and the similarly defined CCCr.

Empirical study shows that equal slice sizes do not make CCC-resolution essentially worse as a price for simplicity of the procedure. Pattern tables show that the mid-sized patterns make the main contribution to the discrimination while the low-sized patterns are equally filled in for competing authors.

A typical table, 1 in [42] (displayed here as Table 3.12), shows that the P-values of the homogeneity testing are minimal for the minimal slice size we dared to consider: 2 KBytes (around half-page). Further studies must show whether even smaller sample size can be chosen without losing validity of SED approximation.

The P-values evaluated from the Normal table rather than from the Student table are a little lower-biased.

The CCC-attribution turns out the same in most case studies when using various popular commercial UCs such as WinZip, BWT, etc., showing attractive empirical invariance of the CCC-method.

LZ-78 was studied in more detail due to its additional application described in the next section. G. Cunningham found that the CCC-resolution of LZ-78 is better when using bytes rather than bits as an alphabet. In the case study of Sect. 3.6.10 the corresponding t-value is 2.88 versus 1.42 for bits.

If $|\overline{CC(T(k))} - \overline{CC(Q(k))}|$ is insignificantfor a slice size then the same turns out true for all larger slice sizes. Plotting CC values over slices serves for checking insignificance of the entropy rate differences between training and query texts and verifies the homogeneity of the style in the training texts itself.

If Mean $CC(Q)$ is significantly different from Mean $CC(T)$ (which can be established using their asymptotic normality in [75], the author of the training text T is very unlikely to be the author of Q. If the **mean** CC**s of** Q **and** T **are not**

Table 3.12 P-value for Venus vs Hero 1 homogeneity under training on Amores, Venus

	Size 10,000	Size 5000	Size 2700	Size 2000
Trained on Amores				
Venus vs Hero 1	0.00973	0.00113	$1 * 10^{-6}$	$2 * 10^{-10}$
Venus vs Hero 2	0.07148	0.03004	0.00421	0.00334
Hero 1 vs Hero2	0.0274	0.0057	$6 * 10^{-6}$	$1 * 10^{-8}$
Trained on Amores + Venus				
Hero 1 vs Hero 2	0.00671	0.00021	$2 * 10^{-7}$	$5 * 10^{-9}$

significantly different, then the smaller the mean $CCC(Q|T)$, the stronger appears the evidence for the similarities in style between two texts, and we expect the mean CCC to be the smallest if trained on the *training text* written by the author of the query text.

Thus the less the mean **Description Length**, the better the evidence for authorship (in accordance with the MDL principle).

Our *assumption that the unconditional complexities of the query and the training texts are approximately equal* is illustrated by the extreme case of a long query text consisting of a repeated identical symbol. In spite of its obvious irrelevance, its CCC is smaller than for any training text.

The program implementing the algorithm in [41] in PERL written by G. Cunningham is published in [45], which is available from the website of SIBIRCON 2010.

3.6.5 Follow-Up Analysis of the Most Contributing Patterns

Given two bodies of text, our goal is to find patterns that occur in one body significantly more than in the other, and to see what linguistic relevance these patterns have. The patterns we examine are those that arise during the course of compression using the LZ-78 algorithm.

LZ-78 prepares the following by-product—in a binary LZ-tree of patterns constructed during compression, we can evaluate cardinalities $n(\nu, A)$ of subtrees with given prefix ν for training text A—this is the cardinality of paths crossing ν or simply the number of vertices with prefix ν.

G. Cunningham wrote an economic code of LZ-tree storage and $n(\nu, A)$ evaluation using the language PERL and the algorithm from [45]. The ν-subtree is called **interesting** if its 't'-value is large for competing training texts A, A'.

$$`t' := \frac{|n(\nu, A) - n(\nu, A'|}{\sqrt{n(\nu, A)(c_1 - n(\nu, A))/c_1 + n(\nu, A')(c_2 - n(\nu, A)')/c_2}}, \tag{3.21}$$

where $c_i, i = 1, 2$, are total pattern cardinalities for A, A', which have been previously well CCC-discriminated. Finally, strings of English letters corresponding to the highly interesting patterns are tabulated.

To generate meaningful statistics, we divide each file F into *slices* $F_1, \ldots, F_n$ of a given number of bytes s. That is, F is the concatenation $F_1 \cup \cdots \cup F_n$. Now we run the LZ-78 algorithm on each slice, making note of the maximal patterns—the entries in the dictionary that are not the prefix of any other dictionary entry. Since we represent the dictionary as a binary tree, these maximal patterns are exactly the leaves of the tree. Then we calculate a 't'-value for each pattern.

The asymptotic independence of $n(\nu, A)$ for disjoint patterns ν and different A is proved in [1] using the 'Poissonization'. This explains the meaning of our 't'. However, the vast abundance of patterns makes evaluation of statistical significance hard. Nevertheless, the tables of interesting patterns may contain unexpected ones for a linguist studying comparative styles of competing authors.

Here is a detailed description of the algorithm:

1. For each of the two files $F_i, i = 1, 2$:
   a. Create a histogram H_i, initializing it so that for each bit pattern ν we have $H_i(\nu) = 0$.
   b. Split the file into slices of the given size.
   c. For each slice:
      i. Run the LZ-78 algorithm on the slice, building a binary tree of patterns which is the dictionary.
      ii. Then, for each maximal pattern ν of the dictionary (leaf of the tree), increment $H_i(\nu)$ by 1.
2. Define $c_i = \sum_\nu H_i(\nu)$.
3. For each bit pattern ν in either H_i:
   a. Define $n_i(\nu) = \sum_\mu H_i(\nu \cup \mu)$, the number of times ν was a maximal pattern or the prefix of a maximal pattern in F_i.
   b. Compute the t value (note slight change of notation compared to (80))

$$t = \frac{n_1(\nu) - n_2(\nu)}{\sqrt{\frac{n_1(\nu)(c_1 - n_1(\nu))}{c_1} + \frac{n_2(\nu)(c_2 - n_2(\nu))}{c_2}}} \tag{3.22}$$

   c. Decode the bit string ν to a character string w (ASCII, UNICODE, etc., depending on the input)
   d. Write ν, w, and t to a file.

3.6.6 Results

Three 100 KB files were randomly generated bit-by-bit, each bit independently of the previous bits. Each bit in file F_0 had a 90 % chance of being 0, while each bit in files F_1 and F_2 had a 10 % chance of being 0 with same entropy rates. When the CCC program was run using 4 KB slices, we found that $t(CCC(F_0, F_2|F_2)) = 147.6$ and $t(CCC(F_1, F_2|F_2)) = 0.6$.

Next, we ran the program to find the well-distinguished maximal patterns between F_0 and F_1.

F_0 **v.** F_2	
Bit pattern	Absolute t-value
11111111	131.27
00000000	131.26
0000000000000000	72.04
1111111111111111	71.55
111111111111111111111111	44.05
000000000000000000000000	43.95
00010000	34.3308297
11101111	34.27392504

Now we look at the CCC and distinguished pattern results for certain groupings of the Federalist Papers. We worked here only with Federalist Papers that are well-attributed. The file H is the concatenation of Federalist Papers 1, 6–9, and 11–13, all of which are well-attributed to Hamilton. The file M_1 is the concatenation of Federalist Papers 10, 14, and 37–40, while M_2 is the concatenation of Federalist Papers 41–46, all of which are well-attributed to Madison. We ran the CCC program using 4KB slices, and we found that $t(CCC(H, M_1|M_1)) = 4.9$, while $t(CCC(M_2, M_1|M_1)) = 0.2$. Below we summarize the well-distinguished maximal patterns, using the same files and 4KB slices.

F_1 **v.** F_2	
Bit pattern	Absolute t-value
011111111111111111111111 111111111111111111111111	3.00
1001111111011111	2.89
1110110111111101	2.89
1011110111111111	2.76
1111111011110101	2.71
1111010110111111	2.71
11011111111111111111110111111111	2.65
11111111011111111111101111111111	2.65

M_1 **v.** H	
English string	Absolute t-value
␣woul	3.87
ould␣	3.78
pon	3.71
ent	3.31
uld␣	3.27
is␣	3.07
de	3.01
atel	3.00
onve	3.00
y the	3.00
tion t	3.00

M_1 **v.** M_2	
English string	Absolute t-value
N	3.68
rei	3.32
Stat	3.32
E	3.29
be␣	3.26
␣State	3.16
ue␣	3.05
he St	3.00

3.6.7 Brief Survey of Micro-Stylometry Tools

We review only statistics of texts which are *semantics-free and grammar-free*. Such attributors are equally applicable to any language, even to the encoded messages which are not yet decoded such as wiretapped terrorist oral communications in possibly unknown or encoded language. However, these methods are not always robust w.r.t. spelling errors and their resolution power may be inferior to semantic attributors.

One obstacle for implementing these methods is the *evolution and enrichment of styles* during professional careers of writers. So, unless we perform an analysis of the stylistic features of authors across time, we can only compare texts written at around the same time.

Also, authors can work in different *literary forms* (for instance, prose and verse) which may have different statistical properties. Therefore, appropriate *preprocessing* and segmentation into homogeneous parts must be applied to the texts to avoid heterogeneity of forms. The plays' segmentation must precede any style comparisons via the CCC-attribution. Misspellings and names need to be removed to preserve consistency. *Annotated texts* (e.g. verses with stressed vowels indicators) can be more useful resources for computer analysis than bare texts. Finally, reliable stylometry analysis should take into account all available information about a query text (e.g. *time of its preparation*).

The pioneering stylometry study [54, 55] was based on histograms of word-length distribution of various authors computed for 5 different text strings of length 1000 words from each author. These papers showed significant difference of these histograms for different languages and also for some different authors (Dickens vs. Mill) using the same language. At the same time, histograms of Dickens were close to those of Thackeray in terms of their statistical variability estimated from repeated samples.

The second paper [55] describes the histograms for the Shakespeare contemporaries commissioned and funded by a rich Bostonian A. Hemminway.[3] This study demonstrated a significant difference of the Shakespeare canon (SC) histogram from those of all (including F. Bacon) contemporaries studied but one, calling attention to the striking practical identity of C. Marlowe's and SC histograms. This identity was shown by evaluating partial histograms for certain portions of the corpora studied and comparing their inter- and intra-deviations. The 'morning star' of Elizabethan poetry and drama C. Marlowe, at his age of 29 years, allegedly perished in May 1593 under extremely suspicious circumstances being on bail from the 'Star Chamber' (English inquisition), and awaiting the imminent death penalty, (see, e.g., [61]). This happened two weeks before the dedication was amended into the already published anonymously submitted poem claiming it to be the very first 'invention' (as it is written in the dedication) of W. Shakespeare.

In an unpublished honors project (available by request), S. Li used a slight modification of Mendenhall's method for attributing popular poem 'Twas the night before Christmas' to H. Livingstone rather than to its official author C. Moore supporting the claim in Foster [22].

Another distinction between the authors is in the numbers of English words they used: 8000 in Bacon's works vs. 31,500 in the SC including 3200 invented ones in the SC which is more than F. Bacon's, B. Jonson's and Chapman's joint contribution. However, G. Kruzhkov (personal communication) studied dynamics of acquiring new words in the SC reinventing the well-known Heaps law. He shows that this **rate** in the SC is not the largest among his contemporaries.

The attribution in [76] of a recently discovered poem 'Shall I die...' to the SC implicitly presumes identity of rates of acquiring new words and forgetting others. Thus, their approach appears questionable.

The attribution based on Zipf and Heaps laws parameter estimation [27, 53] proved to discriminate languages but was shown to have insufficient resolution for discriminating between authors writing in the same language.

Next to mention is the *Naive Bayes* (NB) classifier of [60] developed during their long work generously supported by the Federal Government over the authorship attribution (Madison vs. Hamilton) of certain *Federalist papers*. After

[3]Spelling from [23].

fitting appropriate parametric family of distributions (Poisson or negative binomial), they follow the Bayes rule for odds (*posterior odds is the product of prior odds times the likelihood ratio*), when multiplying the odds: Madison vs. Hamilton, by the sequence of likelihood ratios corresponding to the frequencies of a certain collection of relatively frequent function words, obtaining astronomical odds in favor of Madison. *This classifier presumes independence of function words usage, which is obviously* **unjustified** and 'NB-likelihoods' should not be taken seriously. Among many NB-applications is the attribution of Moliére plays to Corneille [36], attribution of parts of 'Edward III' to SC and Fletcher, and sorting out spam e-mails [15].

Some more popular attributors emerging after NB-approach based on the SVM (see [9]) and modeling language as a Markov Chain of order m (m-MC, see [66]) should be also mentioned. Both are very computationally intensive, depend on choosing many parameters in a non-formalized way which makes their performance evaluation a hard problem.

The chronology of Plato works is studied in [17] with a new attributor based on sequences of long and short among last five vowels. The authors also survey previous approaches to this problem.

Elliott and Valenza [18] is based on around 20-year-long study of several hundred ad hoc attributors ('modes') by many undergraduate students at Californian Claremont McKenna College generously supported by the NSF over many years. Elliott and Valenza [18] gives an informal description of how they chose around fifteen 'best' among attributors for distinguishing SC from other corpora of that time. Elementary combinatorial arguments show that reliability of their choice is questionable. It is also not clear whether the preprocessing design shown to be crucial in previous studies (removing names; comparing works written in around the same time to avoid evolution of style influence, etc.), correct evaluation of statistical significance of multiple decisions, and other important issues **were ignored**. These concerns do not let us assess reliability of their inference, although their *segmentation of plays in SC and of Shakespeare's contemporaries* is useful for future researchers.

Skipping discussion of other attributors, we move directly to the CCC-attributors which demonstrated even better performance *in average* in certain applications (apparently first introduced by Dmitry Khmelev in a not easily reachable Russian Proceedings, reproduced in one of appendices to [35], before its tuning and improvement in [41, 42].

[14] gives a survey of numerous previous approaches: the classification and clustering of text libraries of comparable size using 'similarity metrics' mimicking information distances of [6] inspired by analogy with Kolmogorov complexity (KC) and replacing KC by commercial universal compressors satisfying certain properties. Symmetry of distance was an issue in these papers while **statistical assessment of attribution was completely ignored**, see [5, 37]. Moreover, the preprocessing stage seems to be missing in [14] which may explain they paradoxical claim that L. Tolstoy stays separately in the tree of the classical Russian authors.

They apparently *forgot to remove substantial installments of the French language in Tolstoy* which has an entropy rate different from that of Russian.

The main distinction of our method of all the previous approaches is our compression of *many slices of the query text* enabling an **applied statistical analysis** of their conditional complexities in terms of their location centers and spread. In this way, we can judge about statistical significance of mean CCC-differences similarly to [54, 55]. We show in [42] that **NCD-attributor of [37] fails** in attributing authorship cases, which are **significantly discriminated with our CCC-attributor**.

Since distribution of the CCC over slices was only *empirically* established at that time, we illustrated our results in [42] mainly by convincing histograms and plots showing consistency and approximate normality of the CCC-attributor. The latter was also used there for evaluating the P-value of attribution.

Finally, [21, 33] apply alternative stylometry attributors to works of the 'author' we study in Sect. 3.6.12. Fomenko and Fomenko [21] use a version of [59] attributor (vigorously criticized in [50]) to show implausibility of Sholokhov's authorship of a Nobel prize winning novel. The same method was used by their son to show that the 'History of Russia' by M. V. Lomonosov was falsified by G. Mueller. This claim is difficult to check because texts of that time are of questionable attribution. Kjetsaa and Gustavsson [33] showed that the mean sequence sizes in Sholokhov and his rival in authorship F. Kryukov proposed in [74] are significantly different. The author of [3] agrees with [33] in implausibilty of Kryukov's authorship using literary arguments. The authors of [51, 52] used very elaborate lingua-statistical procedure to conclude that this disputed novel was apparently written by A. Serafimovich.

3.6.8 Attribution of Literary Texts

The asymptotics supporting our approach for large samples by modeling language as a SED process may have dubious accuracy for moderate samples and **should be supported by attributing many literary texts with known authorship**. These are Sects. 3.6.9–3.6.11 and 3.6.13 (to some extent). Other sections deal with cases with unknown or disputed attribution. Sections 3.6.9–3.6.12 were processed by S. Brodsky with winzip, Sects. 3.6.13–3.6.17—by I. Wickramasinghe with the same UC, the first examples in Sect. 3.6.15 were earlier processed by S. Li with BWT, see [41], 3.6.10–3.6.13 were also processed by G. Cunningham with LZ-78.

We use our sliced CCC-attributor, verifying insignificant variability of unconditional complexity and validity of CCC-Normal approximation in every case study.

3.6.9 Two Translations of Shakespeare Sonnets

In his pioneering statistical linguistics works, A.N. Kolmogorov singled out three sources of texts variability—their *information content, form and unconscious author's style*. Since we are only interested in the latter, testing resolution power of attributors for texts with identical content and form is of special interest to us.

Thus around 20 different professional translations of the Shakespeare Sonnets into Russian is a good material for comparing designs of text preprocessing and attributors. Of course, the identity of the information content causes certain statistical *dependence between slices' CCC* which makes the analysis accuracy slightly worse than for independent slices—the t-criterion of the mean 'matched pairs' difference between CCC has the number of degrees of freedom twice less than that for independent slices.

Good compression of the Sonnets' classical translations by Gerbel and Marshak due to a **regular structure** of text (about four times) can at the same time worsen CCC-discrimination for improper preprocessing designs. This was shown by comparing the CCC-attribution for two designs: without and *with preprocessing removal of the regular structure*. The latter has larger t-value, although both are significant.

We used the standard preprocessing design—removing the verse form-based carriage returns, spaces and capital letters, and cutting whole sonnets' file into 70 equally sized slices of size 20,006 bytes was applied.

We compared the inter-CCC $Inter_i$ of each slice **trained on the total alternative translation of all sonnets** with the intra-CCC $Intra_i, i = 1, \ldots, 70$, of the same slice under alternative translation **trained on the remaining part of the same alternative translation**.

The significant correlation between intra- and inter-CCC of slices (Pearson's r is slightly less than 0.3) is due to the identity of the information content in both translations.

The 'matching pairs' t-test between 70 measurements is

$$t = (\bar{Inter} - \bar{Intra})/s_d = 4.69, s_d = StD(Inter - Inter). \tag{3.23}$$

The **unconditional CCr for both translations are insignificantly different and only slightly less than 1/2**.

Thus, the translations by Gerbel and Marshak are firmly CCC-discriminated under standard preprocessing.

The approximate normality is seen from the normal plots of Mean inter- and intra-CCC in Fig. 3.15.

3.6.10 Two Books of Isaiah

A CCC-comparison of the two books Isaiah1 and Isaiah2 from the Bible was suggested by Prof. J. Ziv, the zip-inventor and a former President of the Israel Academy of Sciences, who kindly connected us with Prof. M. Koppel, Bar Ilan University. Prof. Moshe Koppel provided us with the MS-Word files in Hebrew (which were converted by us into txt-files using Hebrew MS-Word routine) and

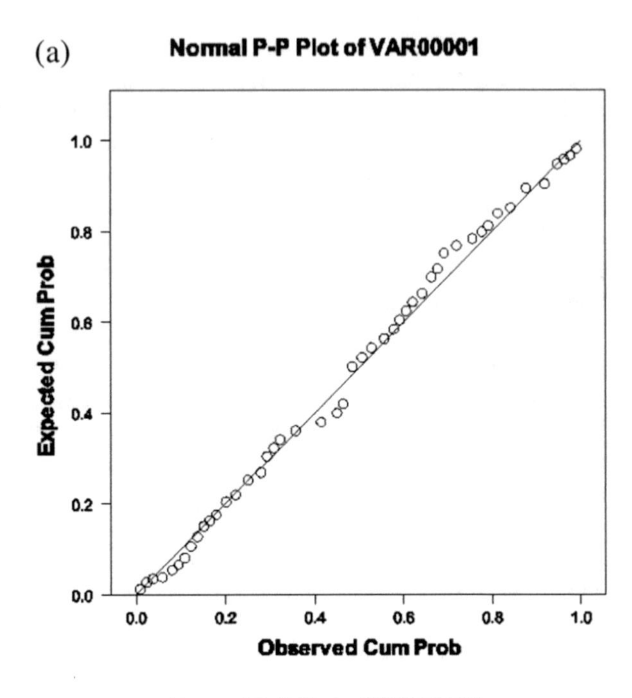

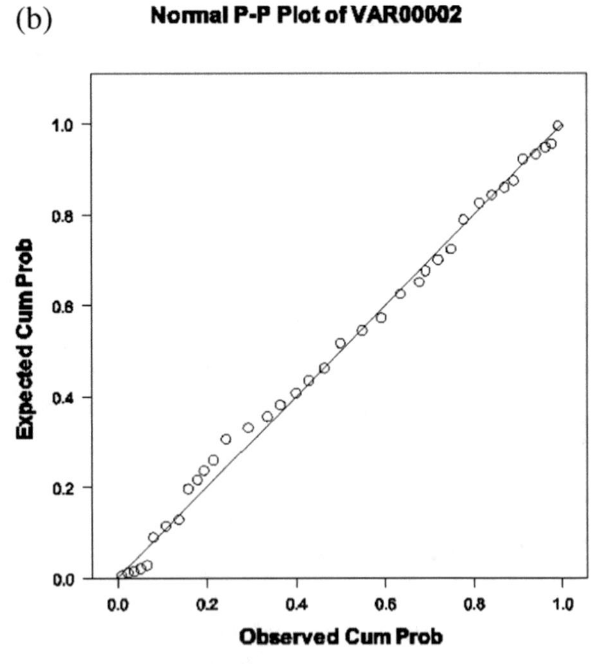

Fig. 3.15 (**a**) Normal Plot: inter-CCC(slices of Gerbel | whole Marshak). (**b**) Normal Plot: intra-CCC(slices of Marshak | remaining Marshak)

informed us that in the 1980s an Israeli named Yehuda Radday did quite a bit of research on the Isaiah question and related Biblical authorship questions.

Our routine preprocessing including removing names and unnecessary punctuation was made by Andrew Michaelson, an alumnus of the Maymonides school, Brookline, MA.

Using the same method as in previous section, G. Cunningham found:

i. the compressed length $|A_c^1|$ of Isaiah1 using LZ78. (201,458 bits when reading octets, 284,863 bits when reading bits instead.)
ii. For each of 21 2Kb-sized slice Q Isaiah1, the compressed length of (Isaiah1 with deleted Q).
iii. For each of 16 2Kb-sized slice P of Isaiah2, the compressed length of (Isaiah1 augmented by P).
iv. The intra-CCCs by subtracting ii. from $|A_c|$ of Isaiah1 and inter-CCC by subtracting $|A_c|$ from iii.
v. The empirical intra and inter means and variances separately.
vi. The t-values evaluated as in the previous section turned out to be 2.88 when reading octets with LZ-78; and 1.42, when reading bits instead. We regard the first one as better corresponding to the techniques used in our other applications.

The Normal CCC-plots are displayed in Fig. 3.16.

3.6.11 Two Novels of the Same Author

The first short novel [10] describes what happened to the author at the end of the twentieth century, while in [11], the same author tries to mimic a language of a boy of twelve, when telling about events in the middle twentieth century. Thus, the author tried to make styles of these two short novels **intentionally different**. Nevertheless, the mean CCC-difference between inter- and intra CCC's for these two works of **the same author** is negligible (t-value is 0.062) in spite of a larger size of two works compared: 144 slices of 2000 bytes. This lack of discrimination (***t*-value grows roughly as the square root of the number of slices, if other parameters are fixed**) shows that the CCC-analysis can successfully resolve authorship problems. The Pearson's $r = 0.058$ is not significant showing practical independence between the inter- and intra-CCC's. The approximate CCC-normality is seen in Fig. 3.17.

3.6.12 Inhomogeneity of an Early Sholokhov's Novel

The first Sholokhov's short novel 'Put'-dorozhen'ka' (abbreviated further as "Put") was published first from September 25, 1925 till May 21, 1925 in the Moscow newspaper 'Young Leninets', issues 93–97, 99, 101–104, 106–114 (*at his alleged age of 20*) and reprinted many times since. With less than 4 years of a primary war time education mostly in a Don village and a brief Rostov 'prodnalog

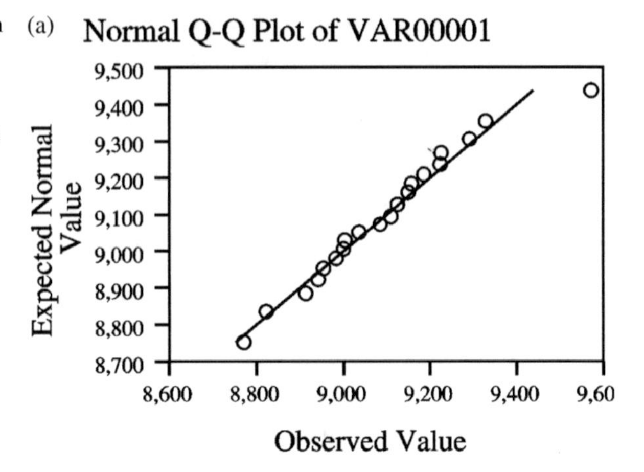

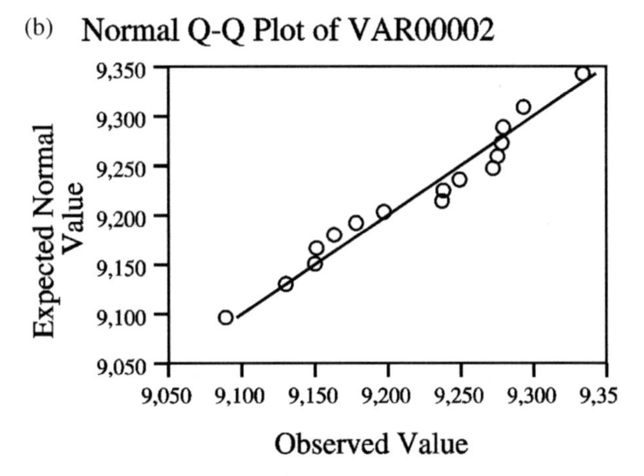

Fig. 3.16 Normal Plots when reading octets with LZ-78: (**a**) intra-CCC(slices of Isaiah1 | remaining Isaiah1); (**b**) inter-CCC(slices of Isaiah 2 | whole Isaiah1)

collecting' accounting courses, he was imprisoned during his subsequent service on the corruption charges for a short time and freed after allegedly forging his age as two years less to be immune from imprisonment. After his release from jail, he soon flees the devastated Don for Moscow in late 1922. There, he was employed for a considerable time by a senior secret police officer Mirumov involved in Stalin's program of '*preparing proletarian writers*'. Mirumov met and 'befriended' M. Sholokhov first most likely during his education in the Rostov 'prodnalog collecting' accounting courses. According to [3], Mirumov might give to A. Serafimovich, a leader of 'RAPP'-the Bolshevik Writers Organization, the manuscripts of a talented dissident Rostov newspaper editor V. Krasnushkin. V. Krasnushkin authored numerous articles and two short novels under the pseudonym Victor Sevsky. He was from the circle of a famous Russian poet Balmont. V. Sevsky was caught and apparently liquidated by the Soviet secret police in Rostov jail (1920). There are reasons to believe that M. Sholokhov was an illegitimate son of A. Serafimovich [51, 52].

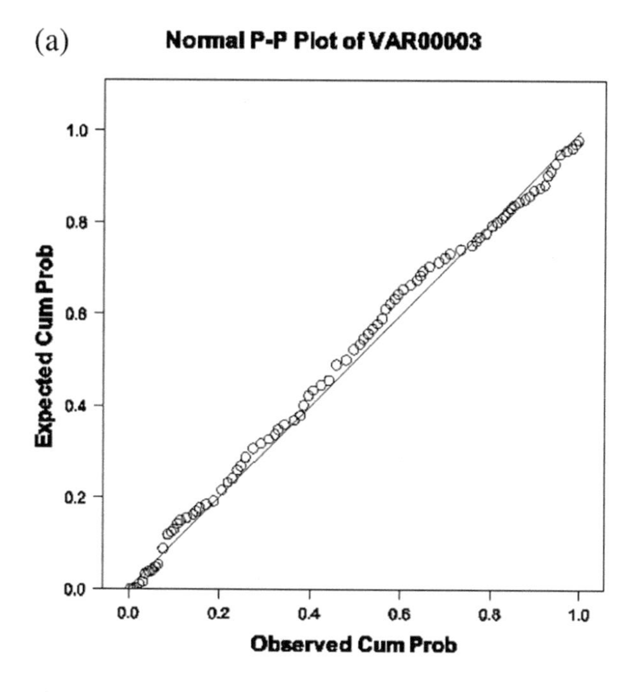

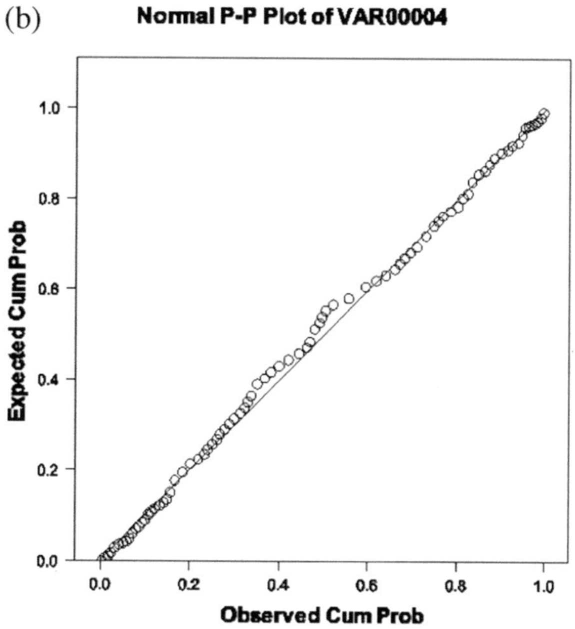

Fig. 3.17 (**a**) Normal Plot: inter-CCC(slices of Brodsky 1 | whole Brodsky 2), (**b**) Normal Plot: intra-CCC(slices of Brodsky 2 | remaining Brodsky 2)

Apparently, A. Serafimovich was involved in reworking anti-Soviet manuscripts into more suitable works allegedly 'written by proletarian writers'. M. Sholokhov publishes the first short story in late 1924. He leaves Moscow for his native village the same year, marries a daughter of a semi-professional writer and stays there for several years (interrupted by his comparatively short visits to Moscow, where he occasionally lives in Mirumov's apartment and dedicates to him his first publications) preparing 'Put', other 'Don stories' and by 1927, the first two parts of the famous huge novel 'Quiet flows the Don' which earned him later the Nobel prize. Any evolution of his style between the first and second parts of "Put" of 12 pages each is extremely unlikely.

After removing names, we partitioned each part into 30 equal slices of 2000 bytes. The Mean Unconditional Complexities ($\bar{CC}$) are statistically the same. The mean 'Leave one out' (intra)-CCC in each part was compared with the mean inter-CCC of each slice trained on another part. Their Standard Deviations are not significantly different. **The difference between Mean inter-CCC and mean intra-CCC turned out to be highly significant (exceeding four its Standard Deviations)**.

Details are as follows: we computed thirty inter-CCC(slice of Part 2|whole Part 1) and thirty intra-CCC(slice of part 1|remaining part 1).

Mean inter CCC is $M_1 = 576.77$ Mean intra CCC is $M_2 = 559.43$, their difference is 17.34, StD (Mean inter CCC)= $s_1 = 2.49$, StD (Mean intra CCC)= $s_2 = 3.50$, finally, the StD of $M_2 - M_1$ is

$$s_d = \sqrt{(s_1^2) + s_2^2)} = 4.30. \tag{3.24}$$

F-ratio < 2 permits using two-sample t-test with test statistic

$$t = (M_2 - M_1)/s_d = 4.03. \tag{3.25}$$

These large t-value and number 58 of degrees of freedom make the corresponding P-value, (i.e. the Student probability of this or higher CCC-deviation) of order 10^{-4}.

Remark We consider inter-CCC of different slices almost independent in our computations which seems reasonable. Intra-CCC may have slight correlation (for example, the sample correlation coefficient between first and last fifteen Part 1 intra-CCC's is 0.156). However, we believe that this correlation would not have a significant impact on the t-value, see [58]. Figure 3.18 shows the approximate normality of the CCC's distribution.

Our two-sample t-test evaluation suggests that the *two parts were written by different authors*. This semantics-free conclusion coincides with that of unpublished sophisticated literary analysis of Bar-Sella.

Let us emphasize that **our analysis and unpublished literary analysis of Bar-Sella are based on different features of the text and thus complement and support each other**.

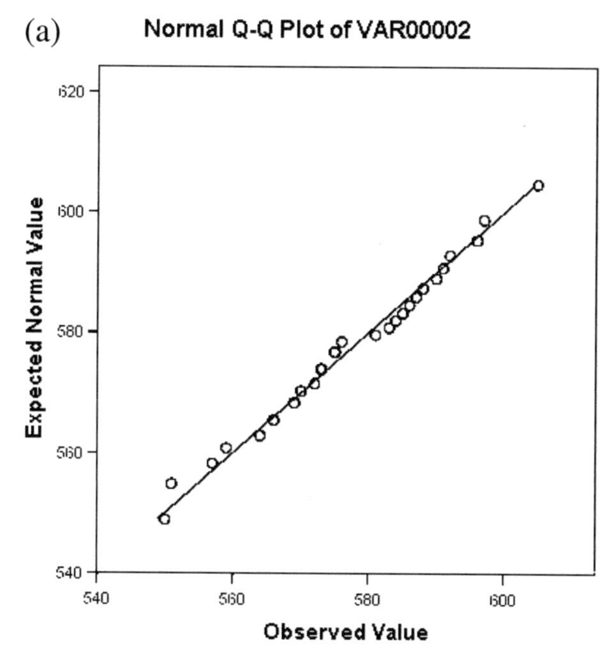

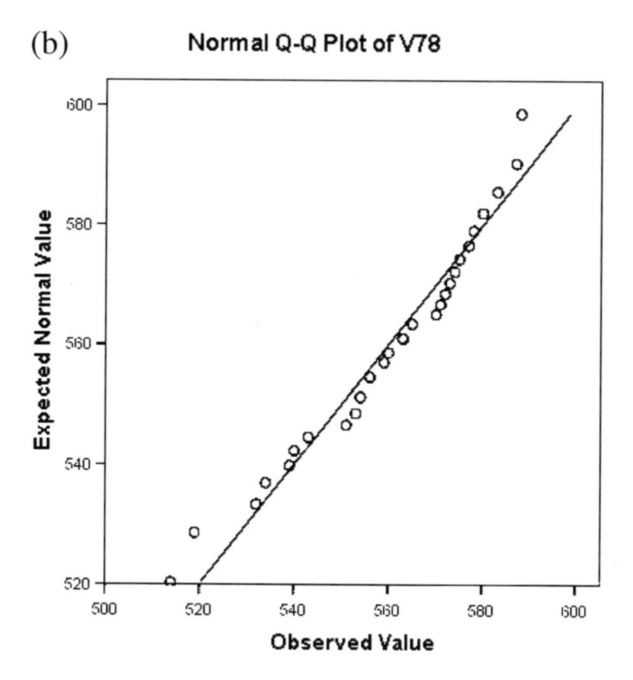

Fig. 3.18 'Put' analysis: (**a**) Normal Plot: inter-CCC(slices of Part 2|whole Part 1), (**b**) Normal Plot: intra-CCC(slices of Part 1|remaining Part 1)

3.6.13 *Attribution of Federalist Papers*

3.6.13.1 The Federalist Papers

The history of the Federalist papers written by Alexander Hamilton, John Jay and James Madison is outlined in our Sect. 3.5.9. Here we answer to the following questions: (1) Does CCC-attribution agree with previous decisions? (2) What slice size is around optimal? (3) What training text size is sufficient?

Answers are: 'yes' on first questions: all Mf and all Df were attributed to Madison, all Hf trained on HF were classified as Hamilton's (2). minimal slice size 2Kb among the tested ones provides the best discrimination (3). Thirteen Mf papers together with Madison's Helvidius papers of the total size around 280Kb is sufficient for significant attribution. Detailed tables are in [77]. **We present a typical tiny sample** of those results.

3.6.13.2 Training on One of Mf-Papers

It was not sufficient for reliable attribution. Namely, Fig. 3.19 shows that trained on one of Mf, the Mf- and Hf-CCC plots are overlapped inside their confidence intervals for all slices sizes.

3.6.13.3 'Leave One Out' Mf as *Training Text*

Four of five Mf-essays (No. 10, 14, 37, 41, 47) were combined leaving one out. Five documents of the size of about 62,000–72,000 bytes after preprocessing were obtained. Our *query texts* are 12 disputed, 5 of the Hamilton essays (No. 07, 09, 11, 30, 70), 2 more of Madison essays (No. 46, 48) as well as the other Madison paper we left out when we combined them as the training set. Figure 3.19 show small variability of mean unconditional complexities and smaller mean CCC's for query Mf than those of Hf.

- We *trained* the compressor separately on each of four-tuples
- We applied the compressor on the concatenated file xy_i, where $x \in M$ and $y \in \{disputed\ papers\}$ and y_i is the ith part of the essay y
- We carried out this study by dividing the *disputed papers* into file sizes of 2000, 3000, and 5000 bytes.
- Mean *CCC* of disputed papers were compared with that for *query Mf- and Hf-texts* by evaluating P-values for the null hypothesis that the mean CCC of the latter are not more than those of a disputed paper.

We display only one typical Table 3 [77] giving the P-values for the two sample t-test: $CCC(Df49|MF_i) \geq CCC(A_j|MF_i)$ trained on five choices of combined four Mf indicated in its columns with A_i indicated in its rows, when the slice sizes of the *query text* are 3 Kbytes.

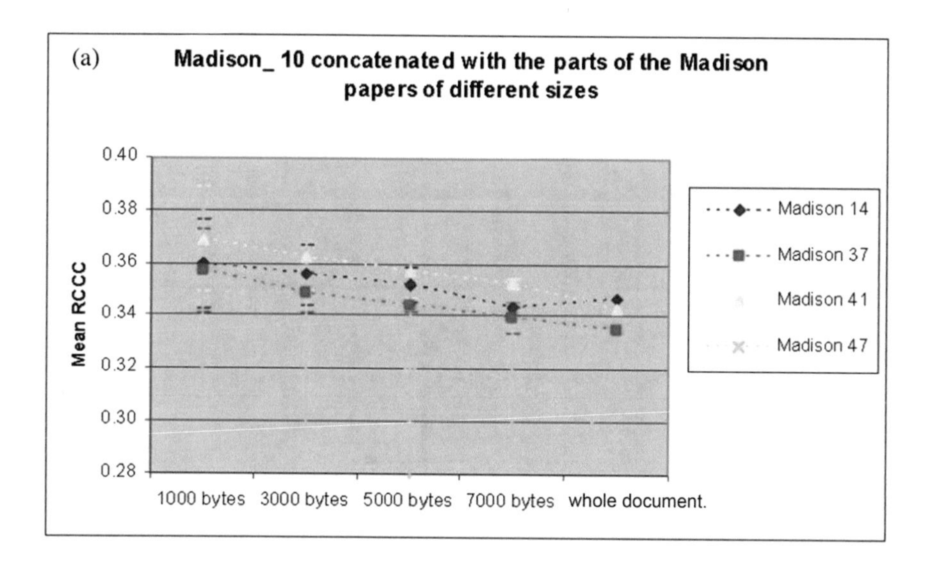

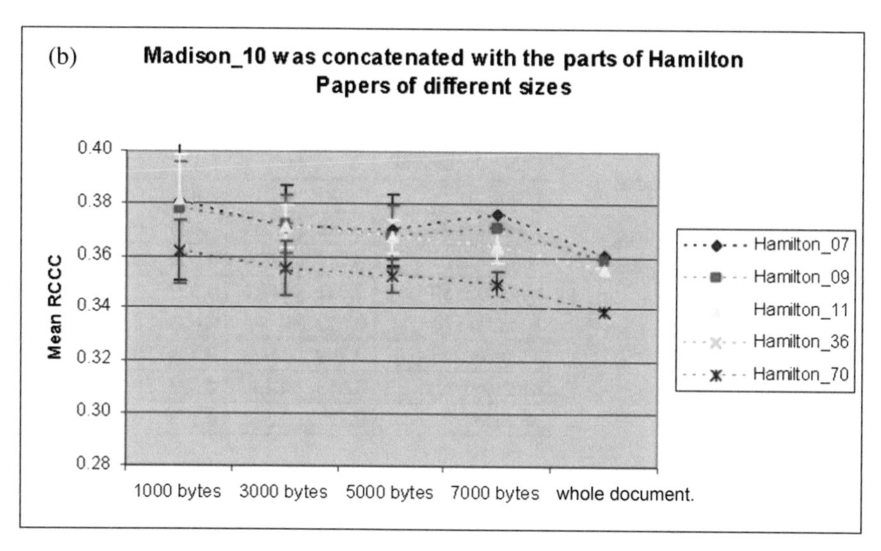

Fig. 3.19 CCC's when trained on one paper

Figure 3.20 shows that *CCC*' empirical distributions are close to Normal.

This and other numerous tables in [77] show insignificant difference of Df's unconditional and Mf's conditional complexities, and attribute most of Df to Madison. Still some exceptions require larger training text, see Table 3.13.

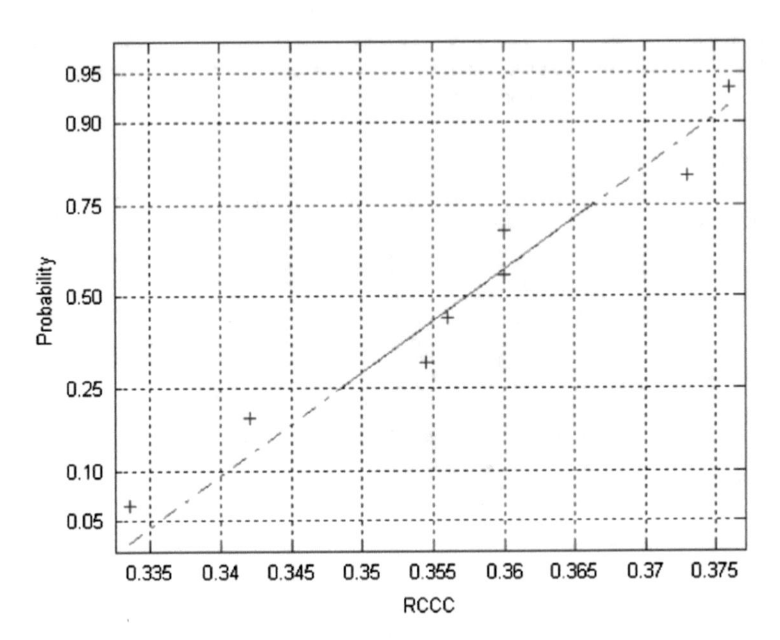

Fig. 3.20 Normal Probability Plot of *CCC* for slices of Hamilton No. 70 of size 2000 bytes trained on Madison essays No. 10, No 14, No. 37 and No. 47

Table 3.13 P-value of the two sample t-test for *disputed paper No. 49*

Other documents	(a)	(b)	(c)	(d)	(e)
Hamilton 07	0.027	0.019	0.018	0.024	0.014
Hamilton 09	0.063	0.065	0.084	0.079	0.077
Hamilton 11	0.010	0.009	0.018	0.021	0.015
Hamilton 30	0.011	0.012	0.027	0.033	0.025
Hamilton 70	0.010	0.053	0.073	0.078	0.069
Madison left-(L)	0.209	0.078	0.200	0.157	0.141
Madison 46	0.215	0.486	0.373	0.371	0.433
Madison 48	0.177	0.342	0.378	0.383	0.400

3.6.13.4 Training on All Mf and Some More Madison Papers

We use here the same technique as before to study attribution for larger *training text*. The following documents were obtained by concatenating all the Mf's leaving only one out. We combined the essays in ascending order of the number of the paper. The federalist papers used are No. 10, No. 14, No. 37–No. 48 which are written by Madison. Sizes of the *training text* varied from 208,000 to 216,500 bytes.

- (a1): Concatenate all except No. 10
- (a2): Concatenate all except No. 14

- (a3): Concatenate all except No. 37
- (a4): Concatenate all except No. 38

and so on.

The following Madison's documents written between 1787–1793 (to avoid evolution influence of the author's style) were used to enlarge *training texts*

- (s): Concatenated four papers out of five (Number 1–4) called "Helvidius papers", written in reply to series by Hamilton called "Pacificus papers" (August 24–September 14, 1793) on executive powers
- (t): Concatenated eight papers from 1791–1792 Congress and republican opposition: (Mad 1: Population and Emigration, *National Gazette*, November 21, 1791), (Mad 2: consolidation, *National Gazette*, December 5, 1791, (Mad 3: Universal Peace, *National Gazette*, February 2, 1792), (Mad 4: Government of the United States, *National Gazette*, February 6, 1792), (Mad 5: Spirit of Governments, *National Gazette*, February 20, 1792), (Mad 6: A Candid State of Parties, *National Gazette*, September 26, 1792), (Mad 7: Fashion, *National Gazette*, March 22, 1792), (Mad 8: Property, *National Gazette*, March 29, 1792).

For collections of total size around 280Kb consisting of (a1)–(a4) and (s) of size 71,010 bytes as *training text*, **all mean *CCC* for Mf's and Df's are significantly lower than that of Hamilton**, making the attribution of Df certain.

3.6.14 The Shakespeare Controversy

Introduction

The controversy concerning authorship of the works ascribed to W. Shakespeare dates back several centuries due to the fact that rare documents related to his life are hard for many to reconcile with his authorship (see e.g. http://shakespeareauthorship.org/).

Many leading figures in poetry, prose and also actors, statesmen, scientists doubt the official authorship version. Around 2000 influential writers, scholars, actors, theater directors, etc. continue to be non-believers and signed the declaration of reasonable doubt to the British government demanding funding for studying the problem, see http://www.doubtaboutwill.org/declaration.

A **bibliography** of material relevant to the controversy that was compiled by Prof. J. Galland in 1947 is about **1500 pages** long (see [23]). A comparable work written today might well be at least several times as large. A substantial part of research moved to the Internet, since publishing works contradicting the official version in academic journals is almost prohibited.

A significant complication for 'heretics' is that they do not agree on the alternative author. The most popular alternative candidate seems to move presently to be Christopher (Kit) Marlowe, sudden death of whom is disputed (see more than a dozen large books appeared in the last few years on this subject). The Hoffman's prize of around 1,000,000 English pounds awaits the person who will prove Marlowe's authorship of substantial part of the SC.

3.6.14.1 CCC-Attribution of Some Elizabethan Poems

We studied the following versions of poems with corrected spelling errors:

- SC: Venus and Adonis (1593), Rape of Lucrece (1594) (we refer to these as Venus and Rape in this study).
- Kit Marlowe's: translation of Ovid's Elegies (Amores).
- Kit Marlowe's: a version of Hero and Leander (Hero 1) both published posthumously in 1598.
- Marlowe's smoother version of Hero and Leander (Hero 2).
- disputed anapest poem 'Shall I die...' earlier attributed in [76].

Kit's translation of Ovid's Elegies (Amores):
http://www2.prestel.co.uk/rey/ovid.htm,
Venus and Adonis (Venus): http://etext.lib.virginia.edu/etcbin/toccer-new2?id=MobVenu.sgm&images=images/modeng&data=/texts/english/\modeng/parsed&tag=public&part=all
Hero and Leander (Hero1):
http://darkwing.uoregon.edu/~rbear/marlowe1.html
Hero and Leander (Hero2):
http://www2.prestel.co.uk/rey/hero.htm
Shall I die, shall I fly:
http://www.shaksper.net/archives/1997/0390.html

These versions with corrected spelling errors in original versions (produced by several publishers in two countries), were recommended to us by British linguist Peter Bull.

Comparatively very long Amores was used as training text which we concatenated with equally-sized slices of the other poems that were used as *query text*. Thus, the size of the training text was not an issue unlike our treatment of *Federalist Papers*. We studied attribution under different sizes of slices, keeping a reasonable number of slices for estimating *StD* of their CCC thanks to large sizes of the poems analyzed. Later we used also the concatenated text of the two poems Amores and Venus as a *training text*.

3.6.14.2 $\overline{CC}$ for the Poems

We calculated the $\overline{CC}$ for each poem divided into slices of various sizes (Figs. 3.21 and 3.22).

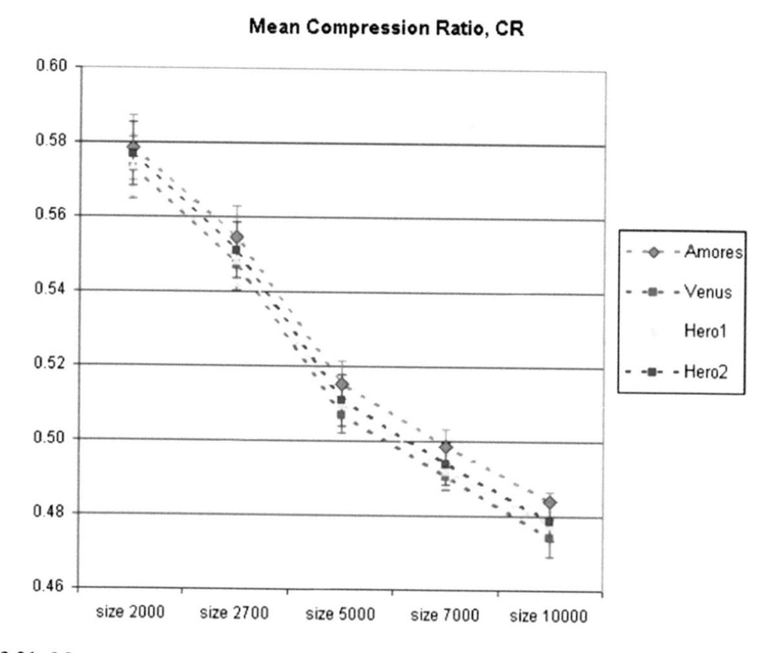

Fig. 3.21 Mean compression ratio *CC* for Amores, Venus, Hero 1 and Hero 2

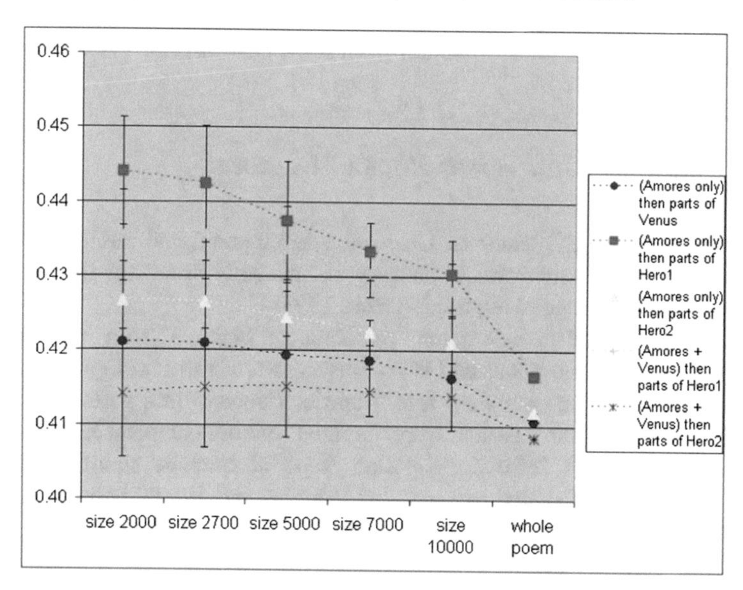

Fig. 3.22 Mean *CCCr* for the concatenated poems

The unconditional complexities for all four poems are **surprisingly close** for any partitioning, which shows an extraordinary consistency of the authors' style. $\overline{CC}$ decreases with the increasing slice size as we discussed in the previous section.

3.6.14.3 Comparison of *CCC* for the Poems

The plots show that in terms of *CCC*, Marlowe's translation of Amores (the first English translation written apparently during his stay in Cambridge around 1585 and published 'posthumously' 5 years later than Venus) helps compress Venus *significantly better than his own Hero and Leander* written allegedly at around the same time as Venus before his alleged 'untimely demise', registered by Marlowe in 1593 and published first separately in 1598 and then (the same year) together with its twice larger continuation ascribed to G. Chapman. Amores was printed in the Netherlands in 1598 and all its copies brought to England were immediately burnt by the orders of Marlowe's deadly foe archbishop Whitgift.

Kit and W. Shakespeare belonged to quite different layers of the society. According to the Wikipedia, master of Cambridge University degree was given to Kit after the unprecedented petition of the Privy Council. He was a high level spy working for two generations of Cecils ruling over Elizabethan England as Prime Ministers. Kit was employed in their covert operations in several countries and for educating a likely successor to the throne. Often, some of his patrons provided him with lodging in their estates. Any interaction with commoner W. Shakespeare associated with a competing theater is not documented and unlikely.

The displayed normal probability plots (Fig. 3.23) support asymptotic normality of the CCC for slices.

Table 3.12 shows extremely low P-values for the two sample *t*-test of a 'natural' hypothesis—$MeanCCC(Hero|Amores) \leq MeanCCC(Venus|Amores)$.

3.6.15 Amores et al. Versus Rape of Lucrece

The second work in SC 'Rape of Lucrece' was prepared and published in haste (1594) thanks to an extraordinary success of unusually erotic for its time Venus which was reprinted around ten times during 1593!

Here we compared three versions of 'Rape of Lucrece' with the poems we studied before: Amores, Venus and Hero 1 using two different compressors winzip and pkzip and dividing our *query text* 'Rape of Lucrece' into parts of size 5000 bytes. Essentially the same results were obtained for both compressors (Fig. 3.24).

We see that Venus helps compressing Rape of Lucrece significantly better than others, the concatenated *training text* 'Amores and Venus' helped even more significantly. Our *query text* 'Rape of Lucrece' was fixed whereas various *training texts* different in size were used for comparison. Whereas Amores is 102,161 bytes, Venus is 51,052 bytes and Hero1 is 33,507 bytes after preprocessing.

One of the explanations of the above results would be that styles of poems following each other almost immediately are closer than those of more timely Amores which eventually was the source for both, while the final editing of Hero took place several years later.

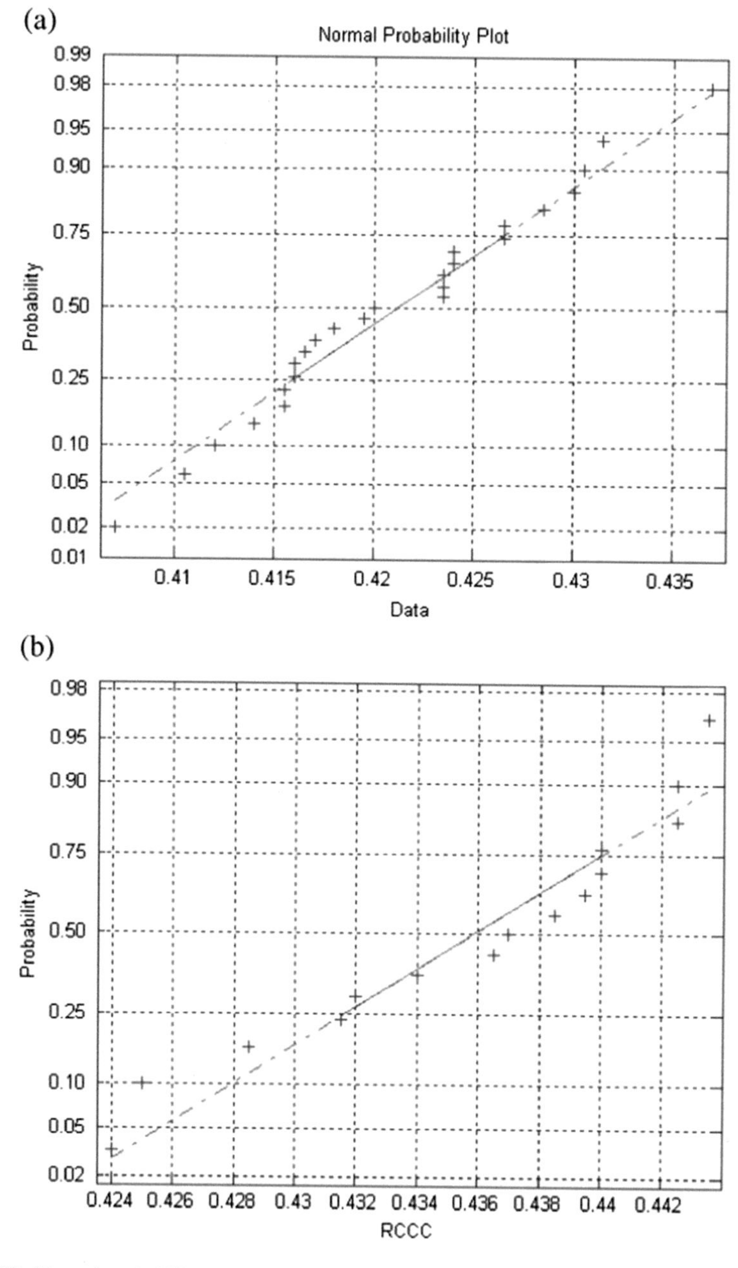

Fig. 3.23 Normal probability plot for *CCCr* of **(a)** Venus trained on Amores, **(b)** Hero 1 trained on the concatenated text of Amores and Venus

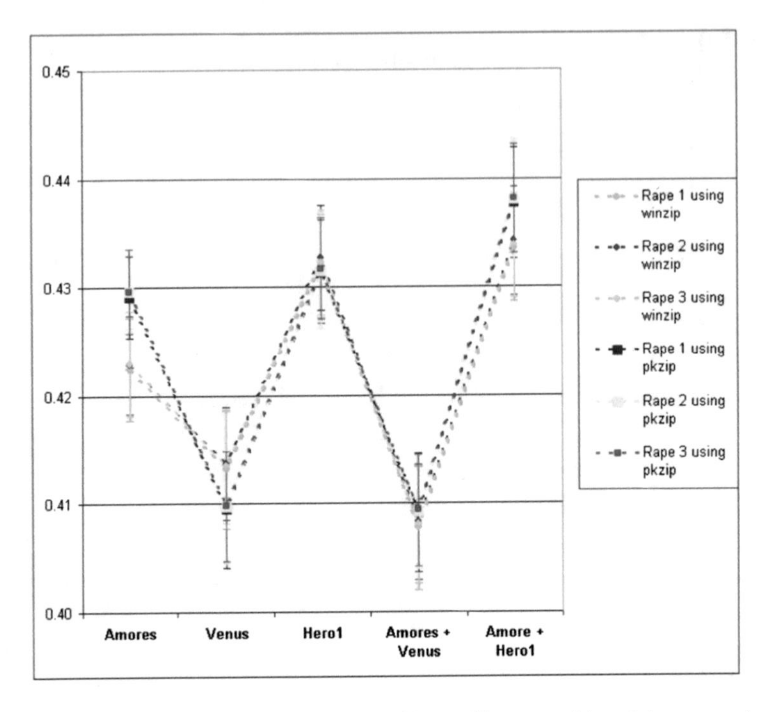

Fig. 3.24 Mean, *StD* of *CCCr* for three versions of Rape of Lucrece with *training texts*—Amores, Venus, Amores and Venus, Amores and Hero1

It is found in [41] that $\overline{CCCc}(Hero2) < \overline{CCCr}(Shall) < \overline{CCCr}(Hero1)$, when trained on 'Amores'. These results make the Marlowe's authorship of both 'Venus and Adonis' and 'Shall I die, shall I fly?' likely.

3.6.16 Hero and Leander Versus Its Continuation

We applied our method to compare the following poems,

- Hero1 (the same as in Sect. 3.6.15) vs HeroChapman 1, a continuation of Hero and Leander written by George Chapman
- Hero 1598 (a version of Hero and Leander) vs HeroChapman 1598, a version of continuation of Hero and Leander written by G. Chapman.

Figures 3.25 and 3.26 show *CCCr*, when query texts are Hero 1, HeroChapman 1, Hero 1598 and HeroChapman 1598 with slice sizes 2.7Kb.

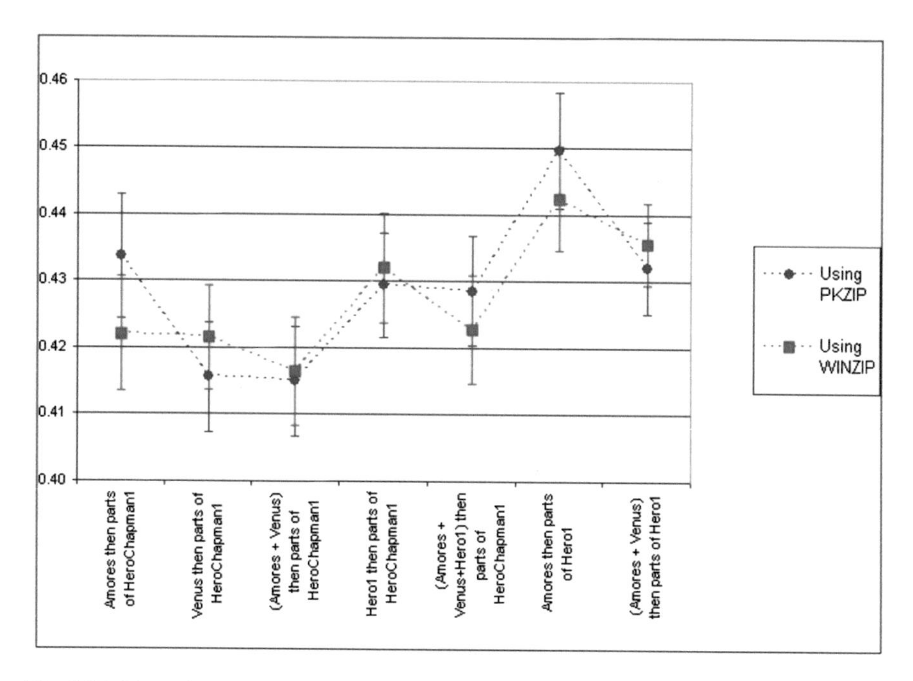

Fig. 3.25 Mean, *StD* of *CCC* for Hero 1 and HeroChapman 1 and some training texts

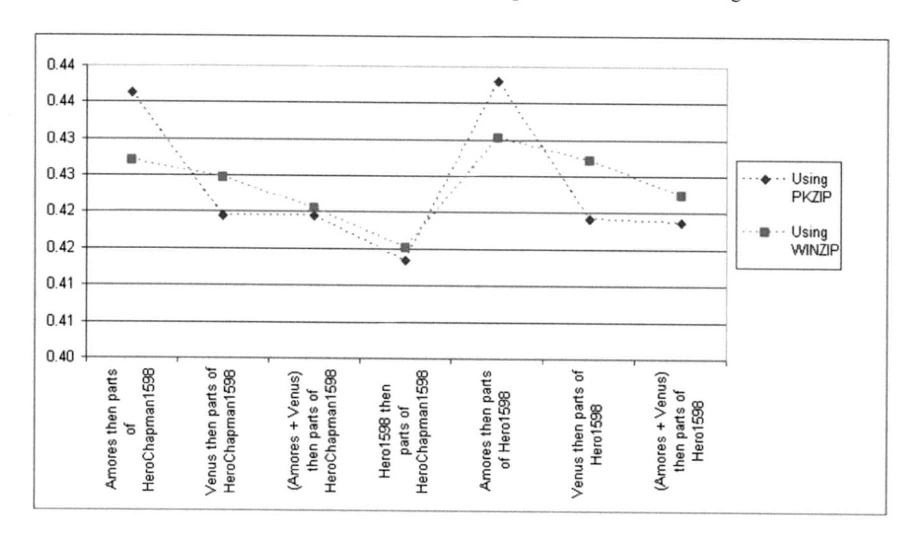

Fig. 3.26 Mean *CCC* taking Hero 1598 and HeroChapman 1598 as disputed poems

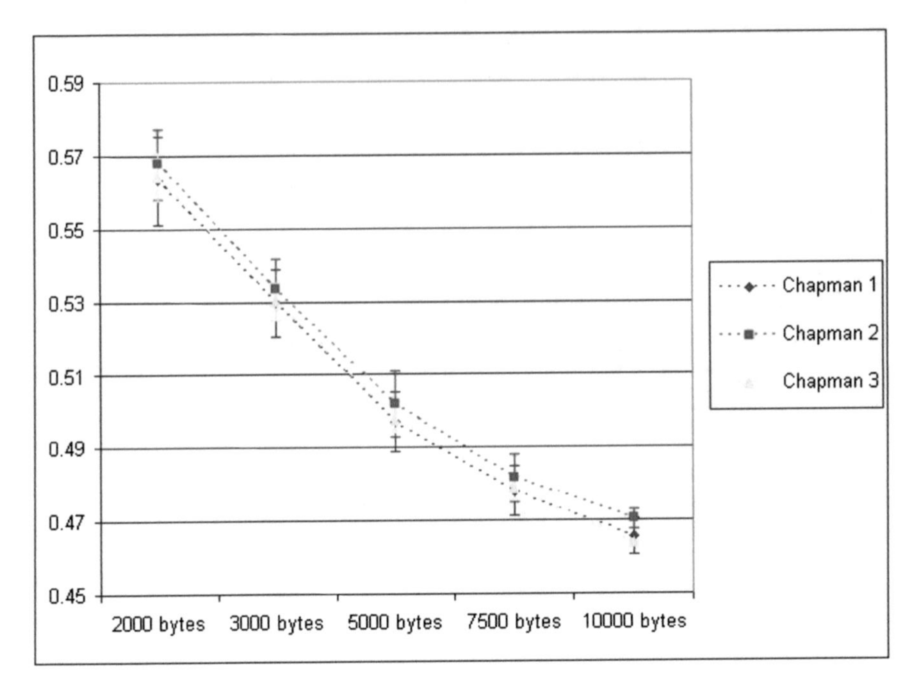

Fig. 3.27 Mean *CCr* for several Chapman's poems

3.6.17 Comparison with Poems Chapman i, $i = 1, 2, 3$

For the style comparison, Peter Bull recommended three poems: Chapman $i = 1, \ldots, 3$, namely 'The Shadow of Night', 'Ovid's Banquet of Sense' and 'The Tears of Peace' written by G. Chapman around the same time (Fig. 3.27). We use also the poems from 5.3 as *training* for 'query' Chapman $i = 1, \ldots, 3$ which were divided into parts of size 3000 bytes. Although the mean *CCC* are lower when both the *training* and the *query text* are firmly Chapman's, a more detailed study is preferable.

Our results do not exclude that G. Chapman helped his friend C. Marlowe in publishing the Kit's 'Hero' by putting his name on its continuation, or Chapman edited both 'Hero' and its continuation. To distinguish between these alternatives, a further more detailed interdisciplinary comparative style analysis of Kit and G. Chapman is desirable.

Remark Malyutov [41] argues that the homogeneity P-value of 154 SC sonnets' fourteen lines versus the presence of anagrammed Marlowe's signatures in first two (four) of them is less than 0.0375 (respectively 2/1000).

Conclusion This **primarily methodological survey** of results obtained by our group in 2004–2011 shows that CCC-testing is efficient in solving attribution problems in several languages. It uses the fundamental Shannon paradigm of modeling sufficiently long LT as stationary ergodic random process and the Kolmogorov-

Rissanen paradigm of statistical decisions based on the MDL principle combining individual and statistical approaches to the texts analysis.

The statistical assessment of our decisions is achieved via averaging which approximates the ensemble means. This is the only situation, (to the best of my knowledge), where the approximation of the incomputable Kolmogorov complexity via commercial UC enables statistically viable and empirically feasible assessment evaluation. An appealing further work would include style attribution of Shakespeare plays or their parts to that of plays written by his contemporaries. Plays segmentation which is crucial for this work was done during the Elliott-Valenza supervized NSF-supported student projects in the Claremont cluster of colleges for many years.

The forthcoming attempt to use the new SCOT training on a cluster of computers with relevant words rather than letters chosen as alphabet during the training is our present goal. It would allow much better follow-up characterization of authors' styles. Success of our statistical approach to LT analysis relies on close collaboration with linguists.

Acknowledgements The SCOT part of Chap. 3 is our joint work with Northeastern University's graduate students Tong Zhang, Paul Grosu, Xin Li and Yi Li [55, 57].

The authors are deeply indebted to Dr. E.A. Haroutunian (Inst. for Informatics and Automat. Problems, Armenian Acad. Sci.) for sending us data and papers on the Radon emissions and seismology, to Dmitry Malioutov and to Libin Cheng (Morgan Stanley, Inc.) for their stimulating discussion of our financial case study.

The CCC-processing work in Sect. 3.5 is due to the hard programming and/or processing work by Slava Brodsky, the Northeastern University's students Gabriel Cunningham, Sufeng Li, Andrew Michaelson, Stefan Savev, Sufeng Li and Irosha Wickramasinghe. Case studies 11.1-4 were processed by Slava Brodsky with useful contribution of G. Cunningham in Sects. 3.6.10–3.6.13. G. Cunningham also wrote a code for finding patterns most contributing to attribution implementing the algorithm [45]. A Northeastern University's student A. Michaelson did preprocessing for Sect. 3.6.10. The statistical simulation of Sect. 3.6 and U-discrimination for IID sources were done respectively by D. Malioutov and S. Savev.

The Springer copyeditor made many remarks on improving language and clarity of our presentation. Edward Dumanis did useful remarks for the first several pages in Chap. 3.

Paul Grosu and Dmitry Malioutov contributed substantially to the LaTeX-processing of this book.

Northeastern University's small grant enabled financial support for G. Cunningham. Peter Bull, Jacob Ziv and Zeev Bar Sella suggested appropriate applications. Moshe Koppel, Z. Bar Sella and Peter Bull sent us original texts for analysis.

References

1. Aldous, D., Shields, P.: A diffusion limit for a class of randomly growing binary trees. Probab. Theory Relat. Fields **79**, 509–542 (1988)
2. Balding, D., Ferrari, P.A., Fraiman, R., Sued, M.: Limit theorems for sequences of random trees. Test **18**(2), 302–315 (2009)
3. Bar-Sella, Z.: Literaturnyi kotlovan—proekt "Pisatel Sholokhov". Russian University for Humanities, Moscow (2005) [in Russian]

4. Bejerano, G.: Automata learning and stochastic modeling for biosequence analysis. Ph.D. dissertation, Hebrew University, Jerusalem (2003)
5. Benedetto, D., Cagliot, E., Loreto, V.: Language trees and zipping. Phys. Rev. Lett. **88**(4), 048702, 28 (2002)
6. Bennet, C.H., Gács, P., Li, M., Vitányi, P.M.B., Zurek, W.: Information distance. IEEE Trans. Inf. Theory **IT-44**(4), 1407–1423 (1998)
7. Bolthausen, E.: The Berry-Esseen theorem for functionals of discrete Markov chains. Z. Wahrsch. Verw. Gebiete **54**, 59–73 (1980)
8. Bolthausen, E.: The Berry-Esseen theorem for strongly mixing Harris recurrent Markov chains. Z. Wahrsch. Verw. Gebiete **60**, 283–289 (1982)
9. Bosch, R., Smith, J.: Separating hyperplanes and authorship of the Federalist Papers. Am. Math. Mon. **105**(7), 601–608 (1998)
10. Brodsky, S.: Bredovy Soup. Limbus Press, Moscow (2004) [in Russian]
11. Slava Brodsky, Funny Children's Stories. Manhattan Academia, New York (2007) [in Russian]
12. Busch, J.R., Ferrari, P.A., Flesia, A.G., Fraiman, R., Grynberg, S.P., Leonardi, F.: Testing statistical hypothesis on random trees and applications to the protein classification problem. Ann. Appl. Stat. **3**(2), 542–563 (2009)
13. Chomsky, N.: Three models for the description of language. IRE Trans. Inf. Theory **2**(3), 113–124 (1956)
14. Cilibrasi, R., Vitányi, P.: Clustering by compression. IEEE Trans. Inf. Theory **51**(4), 1523–1545 (2005)
15. Corney, M.: Analyzing E-mail Text Authorship for Forensic Purposes. Master's Thesis, Queensland Uni. Tech., Australia (2003)
16. Cover, T.M., Thomas, J.A.: Elements of Information Theory, 2nd edn. Wiley, Hoboken (2006)
17. Diaconis, P., Salzman, J.: Projection pursuit for discrete data. Probab. Stat. [Essays in Honor of David A. Freedman, Institute of Mathematical Statistics] **2**, 265–288 (2008)
18. Elliott, W.J., Valenza, R.J.: And then there were none: winning the Shakespeare claimants. Comput. Humanit. **30**, 191–245 (1996)
19. Feller, W.: Introduction to Probability Theory and Its Applications, vol. 1, 3rd edn. Wiley, New York (1968)
20. Fitingof, B.M.: Optimal encoding for unknown and changing statistica of messages. Probl. Inf. Transm. **2**(2), 3–11 (1966)
21. Fomenko, V.P., Fomenko, T.G.: Author's invariant of Russian literary texts. Who was the author of 'Quiet flows the Don?' In (A.T. Fomenko, 'Methods of statistical analysis of historical texts', Appendix 3, **2**, Kraft and Leon, Moscow, 1999) [In Russian]
22. Foster, D.: Author Unknown. H. Holt, New York (2000)
23. Friedman, W., Friedman, E.: The Shakespearean Ciphers Exposed. Cambridge University Press, Cambridge (1957)
24. Galves, A., Loecherbach, E.: Stochastic chains with memory of variable length. In: Festschrift in Honor of Jorma Rissanen on the Occasion of His 75th Birthday, Tampere. TICSP Series No. 38, Tampere Tech. Uni., pp. 117–134 (2008)
25. GARCH toolbox User's guide: The MathWorks, Inc., Natick (2002)
26. Gavish, A., Lempel, L.: Match-length functions for data compression. IEEE Trans. Inf. Theory **42**(5), 1375–1380 (1996)
27. Gelbukh, A., Sidorov, G.: Zipf and Heaps Laws Coefficients Depend on Language. Springer Lecture Notes in Computer Science, vol. 2004, pp. 332–335. Springer, Berlin (2001)
28. Hull, J.C.: Options, Futures, and Other Derivatives, 8th edn. Prentice Hall, Englewood Cliffs (2011)
29. Gut, A., Steinebach, J.: UUDM Report No. 7, Uppsala University Research. Preprints (2002)
30. Hajek, Ya., Shidak, Z.: Theory of Rank Tests. Academic, New York (1967)
31. Haroutunian, E.A., Safarian, I.A., Petrossian, P.A., Nersesian, H.V.: Earthquake precursor identification on the base of statistical analysis of hydrogeochemical time series. Math. Probl. Comput. Sci. **18**, 33–39 (1997)

32. Jensen, J.L.: Asymptotic expansions for strongly mixing Harris recurrent Markov Chains. Scand. J. Stat. **16**(1), 47–63 (1989)
33. Kjetsaa, G., Gustavsson, S.: Authorship of Quiet Don. Solum, Norway (1986)
34. Kolmogorov, A.N.: Three approaches to the quantitative definition of information. Probl. Inf. Transm. **1**, 3–11 (1965)
35. Kukushkina, O., Polikarpov, A., Khmelev, D.: Text authorship attribition using letters and grammatical information. Probl. Inf. Transm. **37**(2), 172–184 (2001)
36. Labbe, D.: Corneille dans l'ombre de Moliére?. Les Impressiones Nouvelles, Paris-Bruxelles (2004)
37. Li, M., Chen, X., Li, X., Ma, B., Vitányi, P.: The similarity metric. IEEE Trans. Inf. Theory **50**(12), 3250–3264 (2004)
38. Liptser, R.Sh., Shiryayev, A.N.: Theory of Martingales. Mathematics and Its Applications Soviet Series, vol. 49. Kluwer Academic, Dordrecht (1989)
39. Mächler, M., Bühlmann, P.: Variable length Markov chains—methodology, computing, and software. J. Comput. Graph. Stat. **13**(2), 435–455 (2004)
40. V. K. Malinovsky, On limit theorems for Harris Markov chains, I. Theory Probab. Appl. (English translation) **31**, 269–285 (1987)
41. Malyutov, M.B.: Review of methods and examples of authorship attribution of texts. Rev. Appl. Ind. Math. **12**(1), 41–77 (2005) [in Russian]
42. Malyutov, M.B., Wickramasinghe, C.I., Li, S.: Conditional Complexity of Compression for Authorship Attribution. SFB 649 Discussion Paper No. 57, Humboldt University, Berlin (2007)
43. Malyutov, M.B.: The MDL-principle in testing homogeneity between styles of literary texts—a review. Rev. Appl. Ind. Math. **17**(3), 239–243 (2010) [in Russian]
44. Malyutov, M.B.: Recovery of sparse active inputs in general systems—a review. In: Proceedings, 2010 IEEE Region 8 International Conference on Computational Technologies in Electrical and Electronics Engineering, SIBIRCON-2010, Irkutsk, vol. 1, pp. 15–23 (2010)
45. Malyutov, M.B., Cunningham, G.: Pattern discovery in LZ-78 texts homogeneity discrimination. In: Proceedings, 2010 IEEE Region 8 International Conference on Computational Technologies in Electrical and Electronics Engineering, SIBIRCON-2010, Irkutsk, vol. 1, pp. 23–28 (2010)
46. Malyutov, M.B.: Search for Active Inputs of a Sparse System—A Review. Springer Lecture Notes in Computer Science, vol. 7777, pp. 478–507. Springer, Berlin (2012)
47. Malyutov, M.B., Zhang, T., Li, X., Li, Y.: Time series homogeneity tests via VLMC training. Inf. Process. **13**(4), 401–414 (2013)
48. Malyutov, M.B., Zhang, T., Grosu, P.: SCOT stationary distribution evaluation for some examples. Inf. Process. **14**(3), 275–283 (2014)
49. Malyutov, M.B., Zhang, T.: Limit theorems for additive functions of SCOT trajectories. Inf. Process. **15**(1), 89–96 (2015)
50. Markov, A.A.: On an application of statistical method. Izvestia Imper. Acad. Sci. Ser. VI **X**(4), 239 (1916) [in Russian]
51. Marusenko, M.A., Piotrowski, R.H., Romanov, Yu.V.: NLP and attribution of pseudonymic texts—who is really the author of 'Quiet Flows the Don'. In: SPECOM 2004: 9th Conference Speech and Computer, Saint-Petersburg, 20–22 September 2004, pp. 423–427
52. Marusenko, M.A., Bessonov, B.A., Bogdanova, L.M., Anikin, M.A., Myasoedova, N.E.: Search for the Lost Author, Attribution Etudes. Philology Department, Sankt Petersburg University (2001) [in Russian]
53. Maslov, V.P., Maslova, T.V.: On Zipf law and rank distributions in linguistics and semiotics. Mat. Zametki **80**(5), 728–732 (2005)
54. Mendenhall, T.A.: The characteristic curves of composition. Science, **11**, 237–249 (1887)
55. Mendenhall, T.A.: A mechanical solution to a literary problem. Pop. Sci. Mon., **60**, 97–105 (1901)
56. Merhav, N.: The MDL principle for piecewise stationary sources. IEEE Trans. Inf. Theory **39**(6), 1962–1967 (1993)
57. Meyn, S.P., Tweedy, R.L.: Markov Chains and Stochastic Stability. Springer, Berlin (1993)

58. Moore, D.S., McCabe, G.P., Craig, B.: Introduction to the Practice of Statistics, 6th edn. Freeman, New York (2008)
59. Morozov, N.A.: Linguistic spectra—stylometry etude. Izvestia Imper. Acad. Sci. Russ. Lang. Ser. **XX**(4), 93–134 (1915) [in Russian]
60. Mosteller, F., Wallace, D.: Inference and Disputed Authorship—The Federalist Papers. Addison-Wesley, Reading (1964)
61. Nicholl, Ch.: The Reckoning. Chicago University Press, Chicago (1992)
62. Reimer, G.M.: Use of soil-gas helium concentrations for earthquake prediction—limitations imposed by diurnal variation. J. Geophys. Res. **85**(B6), 3107–3114 (1980)
63. Rissanen, J.: A universal data compression system. IEEE Trans. Inf. Theory **29**(5), 656–664 (1983)
64. Rissanen, J.: Universal coding, information, prediction and estimation. IEEE Trans. Inf. Theory **30**(4), 629–636 (1984)
65. Rocha, J., Rosella, F., Segura, J.: The Universal Similarity Metric Does Not Detect Domain Similarity, vol. 1, no. 6. Cornell University Library, Ithaca (2006) [arXiv:q-bio.QM/0603007]
66. Rosenfeld, R.: A maximum entropy approach to adaptive statistical language modeling. Comput. Speech Lang. **10**, 187–228 (1996)
67. Ryabko, B.: Twice-universal codes. Probl. Inf. Transm. **20**(3), 24–28 (1984)
68. Ryabko, B., Astola, J.: Universal codes as a basis for time series testing. Stat. Methodol. **3**, 375–397 (2006)
69. Savari, S.: Redundancy of the Lempel-Ziv increment parsing rule. IEEE Trans. Inf. Theory **43**(1), 9–21 (1997)
70. Rassoul-Agha, F., Seppalainen, T.: A Course on Large Deviations with an Introduction to Gibbs Measures. Department of Mathematics, University of Utah, Salt Lake City (2010)
71. Shamir, G., Merhav, N.: Low-complexity sequential lossless coding for piecewise stationary memoryless sources. IEEE Trans. Inf. Theory **45**(5), 1498–1519 (1999)
72. Shannon, C.: A mathematical theory of communication. Bell Syst. Tech. J. **27**, 379–423, 623–656 (1948)
73. Shannon, C.: Communication theory of secrecy systems. Bell Syst. Tech. J. **28**, 656–715 (1949)
74. Solzhenitsyn, A.: Stremya Tikhogo Dona. YMCA Press, Paris (1976) [in Russian]
75. Szpankowski, W.: Average Case Analysis of Algorithms on Sequences. Wiley, New York (2001)
76. Thisted, R., Efron, B.: Did Shakespeare write a newly discovered poem? Biometrika 74, 445–455 (1987)
77. Wickramasinghe, C.I.: The relative conditional complexity of compression for authorship attribution of texts. Ph.D. dissertation, Mathematics Department, Northeastern University (2005)
78. Ziv, J.: On classification and universal data compression. IEEE Trans. Inf. Theory **34**(2), 278–286 (1988)

Printed in the United States
By Bookmasters